AF894045

Philipp Jung

Nonlinear Effects in Superconducting Quantum Interference Meta-Atoms

Experimental Condensed Matter Physics
Band 15

Herausgeber
Physikalisches Institut

Prof. Dr. Hilbert von Löhneysen
Prof. Dr. Alexey Ustinov
Prof. Dr. Georg Weiß
Prof. Dr. Wulf Wulfhekel

Eine Übersicht aller bisher in dieser Schriftenreihe erschienenen Bände finden Sie am Ende des Buchs.

Nonlinear Effects in Superconducting Quantum Interference Meta-Atoms

by
Philipp Jung

KIT Scientific Publishing

Dissertation, Karlsruher Institut für Technologie (KIT)
Fakultät für Physik
Tag der mündlichen Prüfung: 31. Oktober 2014
Referenten: Prof. Dr. Alexey V. Ustinov, Prof. Dr. Steven M. Anlage

Impressum

KIT Scientific Publishing

Karlsruher Institut für Technologie (KIT)
KIT Scientific Publishing
Straße am Forum 2
D-76131 Karlsruhe

KIT Scientific Publishing is a registered trademark of Karlsruhe Institute of Technology. Reprint using the book cover is not allowed.

www.ksp.kit.edu

This document – excluding the cover – is licensed under the Creative Commons Attribution-Share Alike 3.0 DE License (CC BY-SA 3.0 DE): http://creativecommons.org/licenses/by-sa/3.0/de/

The cover page is licensed under the Creative Commons Attribution-No Derivatives 3.0 DE License (CC BY-ND 3.0 DE): http://creativecommons.org/licenses/by-nd/3.0/de/

Print on Demand 2014

ISSN 2191-9925
ISBN 978-3-7315-0294-4
DOI 10.5445/KSP/1000043835

Contents

1. Introduction **1**

2. Background **11**

2.1. Metamaterials & Meta-Atoms 12
2.1.1. Magnetic Metamaterials 12
2.1.2. Planar Magnetic Meta Atoms 13
2.2. Meta-Atoms in Waveguides 16
2.2.1. The Coplanar Waveguide 17
2.2.2. The Rectangular Waveguide 19
2.2.3. The Transmission Line Model 20
2.2.4. Microwave Network Analysis 22
2.3. Superconductivity 25
2.3.1. Flux Quantization in a Superconducting Loop 25
2.3.2. The Josephson Junction 26
2.4. The SQUID meta-atom 29
2.4.1. The Single Junction SQUID 29
2.4.2. Limits of weak and strong driving 32
2.4.3. Intermediate Driving 35

3. Experimental Setup & Procedures **39**

3.1. Samples 40
3.1.1. Fabrication 41
3.1.2. Sample Generations 43
3.2. Microwave setup 48
3.3. Pulsed setup 54
3.4. Modeling 57
3.5. Rectangular Waveguide Setup 61

4. Results 65
4.1. Theoretical analysis . . . 66
4.1.1. Nonlinearity Beyond the Duffing-Oscillator . . . 66
4.1.2. Analysis of the Theoretical Model . . . 70
4.1.3. Comparison with Numerical Simulations . . . 78
4.2. Measurements in the Weak Driving Regime . . . 83
4.2.1. Single SQUIDs . . . 83
4.2.2. One-Dimensional SQUID Arrays . . . 85
4.2.3. Two-Dimensional SQUID Arrays . . . 87
4.3. Multistability and Hysteresis . . . 92
4.4. Pulse-Switching . . . 98
4.4.1. Pulse Protocol . . . 98
4.4.2. Two SQUID Pulse Switching . . . 103
4.4.3. Pulse Switching of a SQUID Array . . . 105
4.4.4. Alternative Excitation Schemes . . . 107

5. Conclusions and Outlook 113

Appendix 125

A. Detailed Calculations 127
A.1. RWA: Complex Fourier Coefficients . . . 127

B. Technical details 131
B.1. Methods . . . 131
B.1.1. Numerical Simulation . . . 131
B.1.2. S-Matrix from Theory (Fig. 4.16) . . . 132
B.1.3. Retrieving SQUID Parameters From rf Measurements . 132
B.2. Instruments . . . 133
B.3. Software . . . 134

1. Introduction

"If we do not take losses into account and regard n, ε, and μ as real numbers, it can be seen [...] that a simultaneous change of the signs of ε and μ has no effect on these relations. This situation can be interpreted in various ways. First, we may admit that the properties of a substance are actually not affected by a simultaneous change of the signs of ε and μ. Second, it might be that for ε and μ to be simultaneously negative contradicts some fundamental laws of nature, and therefore no substance with $\varepsilon < 0$ and $\mu < 0$ can exist. Finally, it could be admitted that substances with negative ε and μ have some properties different from those of substances with positive ε and μ. As we shall see in what follows, the third case is the one that is realized. It must be emphasized that there has not so far been any experiment in which a substance with $\varepsilon < 0$ and $\mu < 0$ could be observed."

Victor Veselago, 1968

Metamaterials & Meta Atoms

The introductory quote is taken from a paper by Victor Veselago published in the late 1960s [Ves68], which is nowadays regarded as the birth of the metamaterial idea. He explores the idea of a material having both a negative electric permittivity ε and magnetic permeability μ and concludes that there are a number of interesting phenomena associated with this situation. At that time, Veselago was looking for and talking about natural materials in which such a situation could be realized. In particular, he pointed out that in order for such a system to be realized, it would require two different mechanisms to achieve negative values for $\varepsilon = \varepsilon_0\varepsilon_\mathrm{r}$ and $\mu = \mu_0\mu_\mathrm{r}$. Here, ε_0 and μ_0 are the vacuum and permittivity and permeability and ε_r and μ_r the relative permittivity and permeability in the medium. For the electric part, the condition is easily achieved in a plasma[1] at frequencies below the plasma resonance. The lack of magnetic monopoles requires a different approach for the magnetic part of the condition which, according to Veselago, could be achieved by the dipole moment of spins.
It was decades later that John Pendry and his group picked up Veselago's idea and proposed a way to achieve this condition with what we today call metamaterials [Pen+96; Pen+99; Pen00]. The novelty of their approach lay in the idea of mimicking the function of atomic spins and the free electron plasma using artificial atoms. There is one central condition that has to be fulfilled in order for these meta-atoms (MAs) to form an effective medium that can then be described by relative parameters ε_r and μ_r: Their size and spacing has to be much smaller than the wavelength in the medium at the frequency of operation. Subsequent experimental implementations by Shelly Schultz, David Smith and their grou [Smi+00; She+01; SSS01; Sch+06] made use of this fact by working in the microwave regime where meta-atoms can be millimeters in size. Their metamaterial implementation consisted of two types of meta-atoms fabricated using easily accessible printed circuit board (PCB) technology. Split-ring-resonators (SRRs) served as resonant magnetic dipoles while an array of metallic wires provided a negative ε_r background. The combination of these elements made it possible to demonstrate a negative-index passband and cloaking properties in the microwave regime.
Since those early days, metamaterials have come a long way. In the wake of promises of tremendous magnitude such as optical cloaking [AE08] and super-

[1] Such as the free electrons in a metal.

resolution imaging [LL12], many other related areas of research have merged with the original idea. Nowadays, metamaterial research covers a multitude of different topics from plasmonic waveguides [NB11] to redirecting sound waves for tsunami-protection [Far+08].
In the more conventional direction that deals with influencing material parameters, one of the main developments has been a constant push towards higher frequencies with the ultimate goal of optical metamaterials in mind.

This development towards higher frequencies, however, brought about a new challenge as losses in resonant, metallic meta-atoms scale inversely with their physical dimensions and degrade or even completely destroy the effects for which those metamaterials were intended [Anl11]. Several routes have been suggested that aim at mending or circumventing this problem such as the idea of compensating losses with gain in the meta-atoms [Boa+11]. The absence of Ohmic losses in superconductors eventually led to the idea of using them to build metamaterials. However, it only became apparent later that they have much more to offer, as we shall see in the next section.

Superconducting Metamaterials

Although superconductors have originally been used in metamaterials as a low-loss alternative to normal metals, they also offer a number of other advantages [Anl11; JUA14]. Fig. 1.1 shows a schematic overview of superconducting circuits that have been proposed or used as superconducting meta-atoms. In the following, I will briefly summarize those advantages and explain some of the new opportunities that arise from effects that are unique to superconductors.

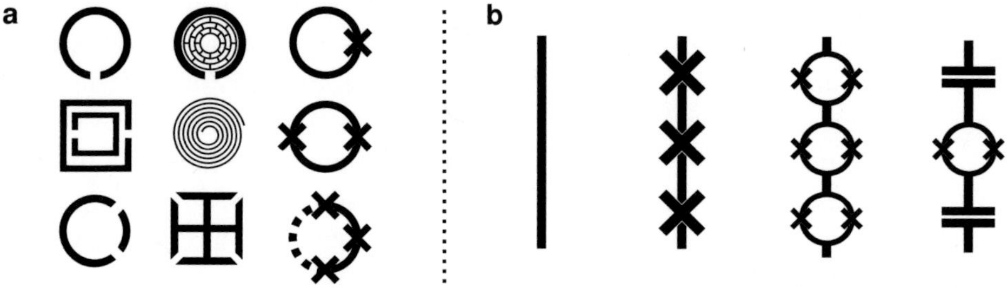

Figure 1.1.: Superconducting meta-atoms and implementations. (The crosses symbolize Josephson Junctions.) **a**, Magnetically coupled meta-atoms. Left column: Different variations of the split-ring resonator. Central column: From top to bottom: Wood-cut type flux quantization MA, spiral. Right column: rf-SQUID, dc-SQUID, and qubit. **b**, Electrically coupled "wire" structures. Upper left: Standard type; provides a negative ε_r below its plasma frequency. Upper right: JJ decorated wire; the plasma edge is tunable by running a current through the wire. Lower left: SQUID decorated wire; The plasma edge is tunable by DC magnetic field. Lower right: SQUID-loaded resonator; provides a negative ε_r above its resonance frequency.

Low Losses and Scalability

The first version of a low-loss superconducting metamaterial consisting of a Nb wire array and thin-film Nb double split-ring resonators (DSSRs) was demonstrated by Ricci et al. [ROA05] in a rectangular waveguide. The authors were able to produce a negative index passband using a geometry similar to the first paper by Smith et al. [Smi+00]. Losses in superconducting, resonant meta-atoms do not scale with dimensions the way they do in normal metallic resonators [RA06; Anl11]. Therefore, meta-atoms made from superconductors can, in principle, be scaled down in size to operate at higher frequencies without losing their low-loss properties. One fundamental upper limit for this scenario is the gap frequency

$$f_{\mathrm{g}} = \frac{2\Delta}{h}. \tag{1.1}$$

Here Δ is the superconducting energy gap. Above f_{g}, the energy of the photons incident on the superconductor is sufficient to break Cooper pairs. It has been shown that up to this frequency, meta-atoms made from low T_{c} materials such as Nb [Jin+10; Eng+13] or NbN [Wu+11; Zha+12] can exhibit quality factors far exceeding those of comparable gold meta-atoms.
On the other hand, meta-atoms made from high T_{c} materials such as YBCO [Gu+10; Sin+13] have a higher gap frequency but suffer from structural or fabricational issues that significantly limit the quality of their resonances.

Tunability

Non-superconducting, tunable resonant circuits often achieve tunability through tunable capacitors while the inductive part of the circuit is constant. At room-temperature, effects such as the voltage-dependent thickness of the depletion zone of a semiconductor PN junction can be used to build tunable capacitors.
The lossless nature of Cooper pair transport, however, leads to the existence of additional (tunable) inductive contributions to the resonance associated with the inertia of the Cooper pairs. In general, these fall into two categories:

First, the kinetic inductance of a superconducting lead L_{kin} depends on the density of Cooper-pairs n_{s} in the superconductor. This, in turn, depends on a number of external parameters such as temperature T, magnetic field H and density of an externally applied current J. All of these have upper critical values T_{c}, H_{c}[2], and J_{c}, respectively, above which the material is no longer superconducting. When one of those critical values is approached from below, n_{s} becomes highly sensitive to variations in any of the parameters. Such effects have been used to demonstrate the tunability of resonances by temperature [RA06; Wu+11] and magnetic field [Ric+07; Jin+10]. This method, however, is usually

[2] H_{c2} for type II superconductors.

associated with increased losses and, in case of strong magnetic fields, entry of Abrikosov vortices into the superconductor.

The second category is the one that this work is primarily concerned with. It involves the *Josephson junction (JJ)* which, in general, is a weak link between two superconducting leads such as a tunneling barrier. Voltage across and current through the junction are governed by the Josephson relations

$$I = I_c \sin\varphi, \tag{1.2a}$$

$$V = \frac{\Phi_0}{2\pi}\frac{\partial\varphi}{\partial t}. \tag{1.2b}$$

Here, $\Phi_0 = \frac{h}{2e}$ is the flux quantum, φ is the gauge-invariant phase difference between the superconducting wave-functions on either side of the junction and I_c is the critical current of the junction. From Eq. (1.2a–1.2b) one can show that for small variations of the phase around a constant value φ_0, the junction behaves as an inductor with an inductance

$$L_j(\varphi_0) = \frac{\Phi_0}{2\pi I_c \cos\varphi_0}, \tag{1.3}$$

called the Josephson inductance.

A superconducting loop interrupted by a Josephson junction is called single junction *superconducting quantum interference devices ((rf-)SQUIDs)*. In this case, L_j becomes tunable through a constant magnetic field normal to the area of the loop. JJs and SQUIDs can be included as tunable elements in superconducting circuits [CBL07; Abd+13; Ovc+13] and also serve as building block for superconducting quantum bits [CW08; DS13] and metamaterials. The latter was proposed by different groups [DCL06; LT07; DCL08; MG10] as a tunable alternative to metamaterials based on conventional designs. This concept, which will be detailed in the following chapters, has only recently been implemented experimentally [Jun+13; But+13a; Tre+13].

Nonlinearity

While the aforementioned effects that cause tunability, all rely on nonlinear properties of the superconductor in some way, the high frequency signals of interest for a metamaterial application are often assumed to be very small. Therefore, simplified concepts (like the kinetic and Josephson inductance mentioned above) can be applied. They facilitate the modeling and integration of superconductors and Josephson junctions into well-established frameworks such as transmission line theory. There are, however, several intriguing effects unique to nonlinear meta-atoms [LSK14] that can consequently only be explored if the small-signal regime is left. Non-superconducting, metallic meta-atoms are intrinsically only very weakly nonlinear, so additional elements (such as varactor diodes or other semiconductor devices) are included into the meta-atom structures to achieve typical nonlinear effects such as bistability [Wan+08; Din+12]. Superconductors and Josephson junctions, on the other hand, can be used to build meta-atoms that exhibit a high degree of intrinsic nonlinearity [LT13; Jun+14; VDS09].

Quantum Effects

Although superconductivity and the Josephson effect are quantum phenomena, their macroscopic nature often allows us to describe circuits containing them using classical dynamic equations. Under the right conditions, however, it is possible to access quantized states in the potential of the respective devices. If said potential is anharmonic enough to address the transition from the ground state to the first excited state individually, the circuit is called a quantum bit (qubit).

Metamaterials made from qubits promise a number of novel effects that have been discussed theoretically [JUA14]. In addition, quantum materials (such as large arrays of qubits) are also a candidate for quantum simulation, since complex (and classically unsolvable) Hamiltonians can be mapped onto these structures. As fabricating large numbers of identical qubits is a very challenging task, experimental progress on this subject has only started very recently [Mac+13; Mac13].

This Work

In this work I investigate the nonlinear properties of rf-SQUID meta-atoms with a focus on multi-stability.

I start in chapter 2 by introducing the theoretical background and the concepts, that are used throughout the rest of the thesis. This includes a brief introduction into the field of metamaterials with a focus on planar magnetic meta-atoms. Then, I discuss how such meta-atoms have to be treated when placed inside waveguides rather than free space. This is important because the different types of waveguides used in the presented experiments have to be taken into account in the analysis. This section also includes an introduction to the transmission line model, which will later be used to model coplanar waveguides containing SQUIDs. The third section of this chapter contains the basics of superconductivity, flux quantization and the Josephson effect. Based on this theoretical foundation, the rf-SQUID meta-atom is introduced in section 4. For this purpose, I elaborate on how the different effects come together and derive the fundamental equation that describes the dynamics of the SQUID as a nonlinear oscillator. I also analyze this equation in the limits of weak and strong driving and show how the intrinsic nonlinearity can be treated in a simplified model.

Chapter 3 is centered around the details of the experiment and the experimental techniques. It opens with a description of the samples and how they are designed and fabricated. Over the course of this work, several generations of samples were produced by different foundries. Of these, I only introduce the ones for which I also show experimental data later on. Their parameters are given in a table at the end of the section. Next, I introduce the setup that was used for most of the measurements including the mounting of the sample and the microwave electronics. Through some modification, this setup can be used for pulsed measurements as outlined in section 3. In section 4, I will show how the transmission line approach can be used to model to our particular setup. The fifth section is dedicated to another experimental setup that was used to measure two-dimensional SQUID arrays using a rectangular waveguide.

The results of this work are presented in chapter 4. It starts with a novel theoretical treatment of the periodically driven rf-SQUID based on the rotating wave approximation. The implications of the model are then analyzed and its results compared to numerical simulations. The model predicts the existence of a dynamic multi-stability under certain driving conditions. This is of special interest in the context of metamaterials, as the SQUID meta-atom responds differently to the microwave depending on which state it occupies. Section 2 contains measurement results of single SQUIDs as well as one- and two-dimensional arrays thereof in the weak driving limit. It also includes a brief discussion of the effects of coupling in two-dimensional SQUID arrays. Then, in section 3, the existence of multi-stability is demonstrated experimentally using hysteretic power sweeps which show a very good agreement with the theory developed at the beginning of this chapter. The idea of switching between the states using microwave pulses is explored in the last section both in experiment and numerical simulations. Using this approach, a SQUID loaded waveguide can be used as an all-optical microwave switch.

Finally, in chapter 5, I conclude this thesis by a summary of the results and an outlook towards future research.

2. Background

"Although this may seem a paradox, all exact science is dominated by the idea of approximation. When a man tells you that he knows the exact truth about anything, you are safe in inferring that he is an inexact man."

Bertrand Russel

Chapter 2 deals with the scientific foundation upon which this work is built. As the introductory quote implies, every aspect outlined in the following is based on some sort of approximation. Seeing that we arrive at conclusions by building theory upon theory, approximation upon approximation, it seems almost unreasonable to believe that the results can indeed be explained by the supposed underlying theories. It is therefore imperative that we understand these approximations, their limitations and their implications for the theories based on them.

This chapter is intended to introduce the reader to the physical concepts and considerations necessary to the understanding of the results. It summarizes information found in literature that is nowadays considered canon. Therefore, instead of citing every concept or formula individually, each section contains at its beginning a short list of textbooks, from which the information is taken.

2.1. Metamaterials & Meta-Atoms

In this section, I will discuss the idea of a magnetic metamaterial, its connection to the present work and the description of planar meta-atoms. As this is mostly limited to the application of the theory of classical electrodynamics in media and mechanics, the reader is referred to textbooks such as [Jac06] or [LL98].

2.1.1. Magnetic Metamaterials

The basic idea of electromagnetic metamaterials was already outlined in the introduction. Here we will focus on their working principle especially for those variants that interact with the magnetic field component of electromagnetic waves. Their behavior may be best understood from Maxwell's macroscopic equations in media.

$$\nabla \cdot \vec{D} = \rho, \tag{2.1a}$$
$$\nabla \cdot \vec{B} = 0, \tag{2.1b}$$
$$\nabla \times \vec{E} + \frac{\partial \vec{B}}{\partial t} = 0, \tag{2.1c}$$
$$\nabla \times \vec{H} - \frac{\partial \vec{D}}{\partial t} = \vec{j}. \tag{2.1d}$$

Here, $\vec{E}$ and $\vec{B}$ are the electric and magnetic field, $\vec{D}$ and $\vec{H}$ are the displacement and magnetizing field for the isotropic, linear case and ρ and $\vec{j}$ are the charges and currents in the medium. The relation between the fields can be expressed through the constitutive relations in terms of the magnetization $\vec{M}$ and polarization $\vec{P}$ of the medium,

$$\vec{D} = \varepsilon_0 \vec{E} + \vec{P} = \varepsilon_0 (1 + \chi_e) \vec{E} = \varepsilon_0 \varepsilon_\mathrm{r} \vec{E}, \tag{2.2a}$$
$$\vec{B} = \mu_0 \left(\vec{H} + \vec{M} \right) = \mu_0 (1 + \chi_m) \vec{H} = \mu_0 \mu_\mathrm{r} \vec{B}. \tag{2.2b}$$

Here, χ_e and χ_m are the electric and magnetic susceptibility, respectively.

In conventional media, the relative parameters ε_r and μ_r are spatial averages over a volume of the material containing a certain number of (microscopic) atoms. Consequently, this macroscopic approach is only valid if the spatial variations of the fields are small over that volume. If we consider light with increasing frequency f, the wavelength of the light in the medium $\lambda = (f\sqrt{\varepsilon\mu})^{-1}$ decreases. Hence, this method ultimately breaks down as λ becomes comparable to the distance between the atoms. Thus, the spatial averaging is usually done over a volume containing at least N^3 atoms[1], with $N \approx 100$.

Let us now limit the discussion to a frequency spectrum for which the variation of the fields is small over length-scales large enough to encompass several electromagnetically active, artificial structures, which we call meta-atoms. If these couple more strongly to the fields than the atoms they are made of, the macroscopic parameters of this meta-material may deviate strongly from that of the host material. In fact, they may be distinctly different from that of any natural material as proven by the demonstration of effects such as the negative index of refraction [SSS01] mentioned earlier.

In the following, we will analyze a single, planar, magnetic meta-atom as well as one- and two-dimensional "materials" made from arrays of such meta-atoms. As the name suggests, we assume that they only exhibit a relevant coupling to the magnetic field component.

2.1.2. Planar Magnetic Meta Atoms

The meta-atom investigated in this thesis is the single junction SQUID. Like many of the other magnetic meta-atoms (cf. Fig. 1.1a), it couples to the magnetic field through currents that are induced in a planar loop structure. As these meta-atoms are usually electrically self-resonant, their behavior can be described by that of a *damped and driven oscillator*. Microscopically, a driving magnetic field $\vec{H}_e$ that is constant over the loop area A and parallel to its surface normal $\vec{A}$, is related to the driving flux in the loop by

$$\Phi_{\text{ext}} = \mu_0 \int \vec{H}_e \, d\vec{A} = \mu_0 |H_e| \, A, \tag{2.3}$$

[1]Later, we will see that under certain conditions this averaging can become one- or two-dimensional with N and N^2 atoms, respectively.

while the total flux Φ is the difference between the external flux and that generated by a counteracting screening current I_{sc} in the loop with inductance L

$$\Phi = \Phi_{\mathrm{ext}} - LI_{sc}. \tag{2.4}$$

Without loss of generality, we will assume all fields to point in z direction and the loop to lie in the xy-plane. The magnetic moment of the meta-atom (which has to have the opposite sign compared to the external field at zero frequency) created by the screening current can be written as

$$\vec{m} = -I_{sc}A\,\hat{e}_z = \frac{A}{L}\left(\Phi - \Phi_{\mathrm{ext}}\right)\hat{e}_z, \tag{2.5}$$

where we used Eq. (2.4). To derive a value for μ_{r} of a volume $V = l^3$ filled with such meta-atoms (with loop radius r and meta-atom periodicity in all directions d), we have to find the magnetization $\vec{M}$ of that volume

$$\vec{M} = \frac{1}{V}\sum_i \vec{m}_i. \tag{2.6}$$

Combing Eq. (2.2b) with Eq. (2.3) and Eq. (2.6) we can calculate the averaged, relative permeability:

$$\begin{aligned}
\mu_{\mathrm{r}} &= 1 + \frac{|\vec{M}|}{|\vec{H}|}\hat{e}_z && \text{(2.7a)}\\
&= 1 + \underbrace{\frac{\mu_0 r}{L}\left(\frac{r}{d}\right)^3}_{\tilde{F}}\underbrace{\left(\frac{\Phi}{\Phi_{\mathrm{ext}}} - 1\right)}_{\chi_\phi} && \text{(2.7b)}\\
&= 1 + \tilde{F}\chi_\phi && \text{(2.7c)}
\end{aligned}$$

Here, I chose the definition of a single meta-atom flux susceptibility χ_ϕ in analogy to the material parameters defined in Eq. (2.2b).

It should be pointed out that the assumptions we are making (i.e. linearity and isotropy of the medium) are in fact not fulfilled for the problems discussed in the following chapters (i.e. the SQUID metamaterial). The relations may still be used, however, if applied correctly. The problem of anisotropy, for example, can be treated in a quasi-isotropic fashion if the structure of the fields is well known (i.e. with light of a specific polarization or in a waveguide). Linearity can also be assumed under some conditions. In our case, the effect of a magnetic metamaterial in a waveguide with propagating waves of one frequency can be mapped to a relative permeability that depends on the frequency ω and amplitude $|\vec{H}_0|$ of the incident wave ($\mu_r \equiv \mu_r(\omega, |\vec{H}_0|)$).

2.2. Meta-Atoms in Waveguides

As mentioned in the previous section, I am going to present theories and experiments that concern the interaction between meta-atoms and electromagnetic waves, and more specifically, their magnetic field component. In such experiments, however, meta-atoms are usually not isolated structures interacting with plane waves. Instead, the waves propagate in waveguides that confine the fields in one or two dimensions. Thus, in order to understand the coupling between wave and artificial atom, it is essential to know the type and field distribution of the propagating mode in the waveguide. In this section, I will therefore briefly introduce two types of waveguides that were used for the presented measurements, the transmission line model, and the concept of microwave network analysis. The latter provides the framework for a simplified treatment of meta-atoms in waveguides. Further information can be found in various microwave textbooks such as [Poz05; Col91].

Placing the meta-atoms in waveguides offers several distinct advantages for our application. First, the local field structure is usually well-defined and, in case of low order modes, usually relatively simple. This is particularly important because almost all metamaterials are anisotropic. Second, the confinement leads to an enhanced coupling between fields and meta-atoms. Third, for microwaves, most radiation sources, detectors and other components already use waveguides in form of coaxial cables. Transitioning between those and other waveguides is much easier and more efficient than coupling to free space and back into a waveguide. To ensure low reflections at transitions between two waveguides, their impedances have to be matched[2]. Fourth, the physical size of waveguides in the confined dimensions is on the order of the wavelength, which makes them comparatively compact and simplifies their integration into a cryogenic setup.

[2]The wave impedance of a material with constitutive parameters ϵ and μ is $Z_\mathrm{w} = \sqrt{\mu/\epsilon}$. More about impedance matching can be found in subsection 2.2.3

2.2.1. The Coplanar Waveguide

Superconducting electronic structures, such as SQUIDs, are usually fabricated on-chip using thin films and lithographic techniques also known from semiconductor chips. Waveguides used with such elements are often fabricated using the same process and are themselves planar structures such as microstrips, slotlines or coplanar waveguides (CPWs).

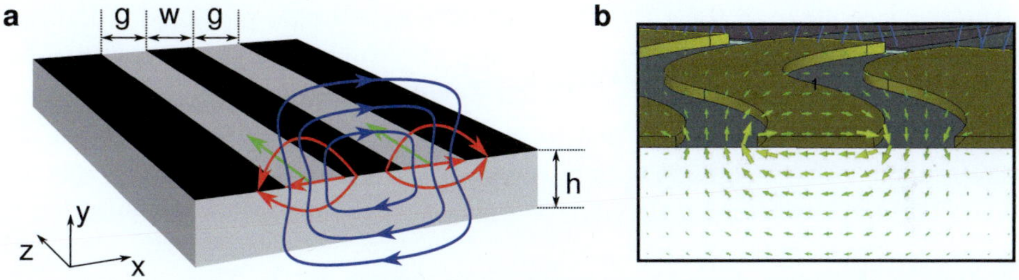

Figure 2.1.: **a**, Sketch of a coplanar waveguide. A metallic layer (black) forms a coplanar structure on top of a dielectric substrate (gray) of height h. The width of the central conductor and gap are shown as w and g, respectively. A microwave, propagating in z direction in the qTEM mode (indicated by the green arrows) is shown with the associated electric (red) and magnetic (blue) field lines. Note that the magnetic field is perpendicular to the substrate surface in the gaps between central conductor and ground plane. **b**, Magnetic field (arrows) of a CPW structure in the qTEM mode calculated using finite element software. The picture shows a cut through the structure perpendicular to the propagation direction with some of the waveguide visible in the top half and the (white) inside of the dielectric in the bottom half.

The latter approach was chosen for the presented work as it offers the strongest coupling between the waveguide and the magnetic meta-atom without electrical contact. The CPW consists of a planar metallic structure[3] on top of a dielectric substrate (cf. Fig. 2.1a). A central conductor of width w is separated from two ground planes by a gap of width g. The ground planes are usually several times wider than $w + 2g$. Being a two-conductor waveguide, the CPW supports a quasi transverse electromagnetic (qTEM) mode starting from zero frequency.

[3]The thickness of the metallic layer is usually much smaller than all the other occurring characteristic length scales.

In this mode, the field structure is particularly favorable for the present experiment. In the gaps between central conductor and ground planes, the magnetic field component is perpendicular to the chip surface, while the electric field is parallel to it. Thus, a flat magnetic meta-atom placed in the gap will not only see the maximum projection of the field onto its surface normal, it will also profit from an increased coupling as the magnetic field lines are concentrated into the gap.

Another advantage of the CPW in the present experiment is related to the fact that it supports a mode at zero frequency. By running a direct current (dc) through the central conductor of the waveguide, it is possible to apply a static magnetic field to the meta-atoms located in the gap without the need for an external coil. This magnetic field bias will be of importance later on, as it is the main tuning-parameter for the SQUID meta-atoms presented in this work.

Naturally, there are also some drawbacks to this approach. The semi-open nature of the waveguide leaves the structure open to couple to modes in its environment. This is especially true if a strong impedance mismatch is introduced into the CPW in which case it acts as an antenna, radiating into the surrounding space [But14].

Impedance matching in this case can be done using analytical expressions such as the ones found in [Sim01]. In practice, however, geometries are often more complicated than the idealized example seen in Fig. 2.1a so that it is usually more convenient to use numerical methods. (See appendix B.3 for details.) Nevertheless, I will discuss some aspect of the simplified model that we will use again later.

Consider a wave propagating in a CPW along the $+z$ direction. In contrast to a wave in vacuum, electric and magnetic fields are now present in two different media (namely, the dielectric substrate and the vacuum, gas or fluid above the waveguide). The properties of the wave (such as its phase velocity $v_{\mathrm{ph}} = c/\sqrt{\varepsilon_{\mathrm{r}}\mu_{\mathrm{r}}}$) are then determined by *effective parameters*. For a CPW with an infinitely thick dielectric in vacuum, for example, the effective, relative permittivity is $\epsilon_{\mathrm{r,eff}} = (\varepsilon_{\mathrm{r}} + 1)/2$ assuming ε_{r} is the relative permittivity of the substrate which occupies the lower half-space while the upper half is is vacuum (with $\varepsilon_{\mathrm{r}} = 1$). Later on, a similar concept will be used to derive a effective, relative permeability resulting from the presence of a metamaterial.

2.2.2. The Rectangular Waveguide

While the coplanar waveguide offers a solid basis for the investigation of single meta-atoms or chains thereof, it is hardly an ideal choice for large-area, two-dimensional arrays or even three-dimensional metamaterials due to its geometry and field distribution. An alternative that has been shown to work well for metamaterials made from superconducting split-ring resonators [ROA05; Ric+07] is the rectangular waveguide. As shown in Fig. 2.2, it is a tube with a rectangular cross section made entirely from metal. The interior can be filled with a dielectric but for our purposes can be considered empty, initially. As it is a one-conductor waveguide, it does not support a mode at zero frequency. Instead it supports a number of modes, the lowest of which is the first transverse electric mode (TE_{10}) which will be the one used for the presented experiments. The frequency range between the lowest mode and the second mode (TE_{20}) sets the limit of operation for this waveguide if we want to ensure that only one mode can propagate.

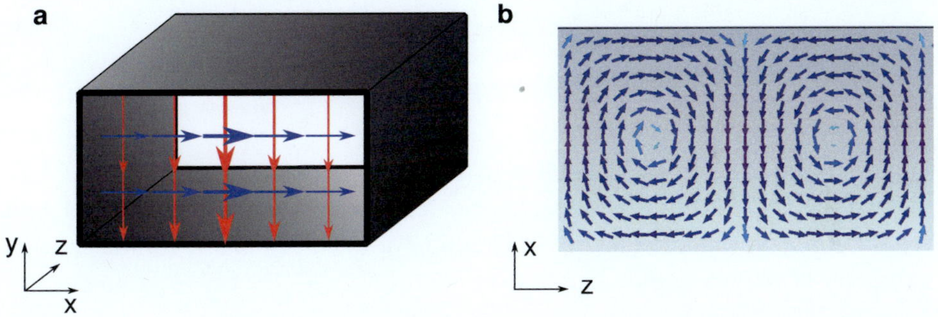

Figure 2.2.: **a**, Sketch of a rectangular waveguide. The fields for a microwave, propagating in z direction in the TE_{10} mode, are shown as red and blue arrows corresponding to electric, and magnetic field vectors, respectively. Their width indicates the strength of the fields. Note that this is a projection onto the x-y plane and the magnetic field also has components in z direction as shown in **b**. **b**, Magnetic field (arrows) of a rectangular waveguide in the TE_{10} mode calculated using finite element software. The picture shows a central cut through the structure perpendicular to the y axis.

In this type of waveguide, modes, field distributions and impedances can safely be calculated using analytic expressions (i.e. found in [Poz05]) as the real devices are usually very well described by the formulas due to their simple geometry. This setup can hold metamaterials with cm^3-size dimensions but suffers from a few drawbacks for our cause: A dc magnetic field has to be applied externally (i.e. by a coil situated outside the waveguide). It also occupies more physical space, which can be a problem in a cryogenic system and mounting the samples can be challenging as shall be explained in more detail in section 3.5.

2.2.3. The Transmission Line Model

A very useful approach that can be used to model waveguides in terms of circuit theory is the transmission line model. It states that in the TEM mode of any two-conductor waveguide, we describe a electromagnetic wave defined by $\vec{E}$ and $\vec{H}$ by a dual model defined by voltage and current. The *telegrapher's equations*

$$\frac{\partial V'(z,t)}{\partial z} = -R'\,I'(z,t) - L'\frac{\partial I'(z,t)}{\partial t}, \tag{2.8a}$$

$$\frac{\partial I'(z,t)}{\partial z} = -G'\,V'(z,t) - C'\frac{\partial V'(z,t)}{\partial t}, \tag{2.8b}$$

take the place of Maxwell's equation in this description. The quantities L',C',R' and G' are the inductance, capacitance, resistance and conductance per unit length of the waveguide (cf. Fig. 2.3) and have to be calculated for the specific geometry of the waveguide in question.

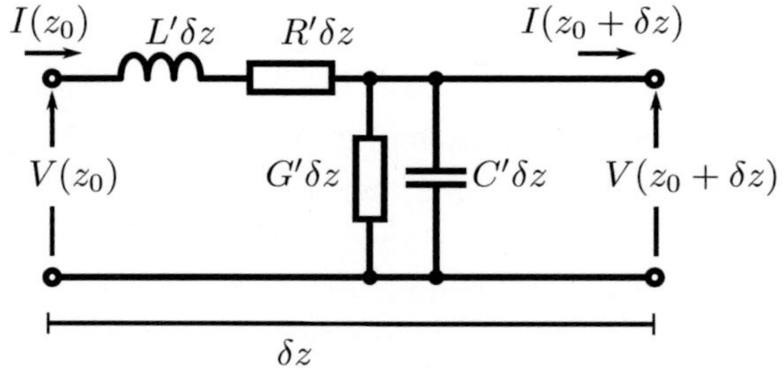

Figure 2.3.: Schematic representation of the transmission line model for a waveguide section of length δz. The impedance of the transmission line (Eq. (2.11)) was derived for the case $\lim_{\delta z \to 0}$

In the following, we will use an assumption that is trivial in case of "normal" network elements such as inductors, capacitors and resistors but becomes non-trivial if we introduce nonlinear elements into the circuit. For now, we can claim that when we apply a wave of only one frequency ω to the system, all voltages and currents can be written using the same time dependence such that

$$V'(z,t) = V(z) \cdot e^{i\omega t} + c.c.\,, \tag{2.9a}$$
$$I'(z,t) = I(z) \cdot e^{i\omega t} + c.c.\,. \tag{2.9b}$$

Later, I will show that for our particular (nonlinear) system, this assumption can still be retained. Under this assumption Eqns. 2.8a and 2.8b simplify to

$$\frac{dV(z)}{dz} = -(R' + i\omega L')I(z) \tag{2.10a}$$
$$\frac{dI(z)}{dz} = -(G' + i\omega C')V(z), \tag{2.10b}$$

and the *line impedance* of the transmission line can be defined as

$$Z_{\mathrm{TL}} = \sqrt{\frac{R' + i\omega L'}{G' + i\omega C'}} \simeq \sqrt{\frac{L'}{C'}}. \tag{2.11}$$

Here, the last approximation is true if the losses are negligible. In practice, the impedance of the waveguide containing the metamaterial has to be matched to that of the other waveguides in the setup (i.e. coaxial cables[4]). The reason for this is that the voltage reflection at the interface between a transmission line with line impedance Z_1 to another with Z_2 is

$$\Gamma = \frac{Z_2 - Z_1}{Z_2 + Z_1}. \tag{2.12}$$

Consequently, reflections can only be avoided if $Z_1 = Z_2$. The main reason why avoiding reflections is so important, is that they lead to standing waves which, for our case, can be considered to be unwanted, parasitic effects.

[4]In the context of this work, the system impedance, to which all components have to be matched, is 50 Ω.

2.2.4. Microwave Network Analysis

So far, I have introduced a framework that allows us to treat a wave, traveling in a waveguide, using an equivalent electric circuit description. Next, I will introduce a method that will allow us to model the scatting properties (i.e. reflection and transmission) of a transmission line loaded with meta-atoms. The resulting scattering matrix S is a quantity that can be measured directly using, for example, a vector network analyzer.

For an electrical network with N ports, we can define the impedance matrix Z with elements

$$Z_{ij} = \left.\frac{V_i}{I_j}\right|_{I_k=0\,\forall\,k\neq j} . \tag{2.13}$$

Here, V_n and I_n refer to the voltages and currents at port $i = 1..N$, respectively. The condition $I_k = 0\,\forall\,k \neq j$ means that when driving port j with a current I_j, all other ports are left open. We are also implicitly making the same assumption that was already used in Eqns. 2.9a and 2.9b.

The voltages and currents can be written as the amplitudes of two waves, one propagating into and one propagating out of the port.

$$V_n = V_n^+ + V_n^-, \tag{2.14a}$$

$$I_n = I_n^+ - I_n^-. \tag{2.14b}$$

The elements of the scattering matrix can then be defined as

$$S_{ij} = \left.\frac{V_i^-}{V_j^+}\right|_{V_k^+=0\,\forall\,k\neq j} . \tag{2.15}$$

Each complex element relates the voltage amplitude and phase of the wave leaving port i of the device to that of the wave going into the device at port j. The condition $V_k^+ = 0\ \forall\ k \neq j$ means that all ports other than the driven port j are terminated with a matched load. The scattering matrix can be expressed in terms of the impedance matrix through the relation

$$S = (Z - Z_0\mathbb{1})(Z + Z_0\mathbb{1})^{-1}, \tag{2.16}$$

where $\mathbb{1}$ is the identity matrix of rank N assuming the system impedance is Z_0 at all ports.

For $N = 1$, Eq. (2.16) reduces to Eq. (2.12) as it should. In this experiment, however, we only deal with two-port networks ($N = 2$), so the transmission from port $1 \rightarrow 2$ and $2 \rightarrow 1$ can be identified as S_{21} and S_{12}, respectively, while reflections on ports 1 and 2 are S_{11} and S_{22}.

Although the magnitude of each of the elements of the scattering matrix $|S_{ij}|$ is the ratio between incoming and outgoing voltages, the value is often given in dB. In this case $|S_{ij}|$ [dB] is the (logarithmic) ratio between the power of the incoming and outgoing waves. The conversion between the two is

$$|S_{ij}|\,[\mathrm{dB}] = 20 \cdot \log_{10}(|S_{ij}|). \tag{2.17}$$

Finding the Z matrix: ABCD Parameters

The model outlined in the last subsection relies on the fact that we already know the impedance matrix Z of the scatterer for which we want to calculate the S matrix. If we know the circuit representation of the device, Z can be found replacing the elements of the circuit with their impedances and applying Kirchhoff's laws. Although this is in principle a trivial task, the results may become quite unintuitive for large networks. An alternative of deriving the impedance matrix for $N = 2$ is the method of ABCD parameters. Together, they form a matrix $\tilde{A}$ which relates currents and voltages at ports 1 to those on 2:

$$\begin{pmatrix} V_1 \\ I_1 \end{pmatrix} = \underbrace{\begin{pmatrix} A & B \\ C & D \end{pmatrix}}_{\tilde{A}} \begin{pmatrix} V_2 \\ I_2 \end{pmatrix}. \tag{2.18}$$

This way, the combined matrix $\tilde{A}_c$ of a series of N networks that are inserted into the transmission line in series can be written as the product of the individual matrices $\tilde{A}_n$,

$$\tilde{A}_c = \prod_{n=1}^{N} \tilde{A}_n. \tag{2.19}$$

In many cases, this approach can be used to split up a complicated network into trivial subnetworks, determine their $\tilde{A}_n$ matrices and calculate the combined matrix $\tilde{A}_c$.

Many of the trivial matrices can also be found in literature such as [Poz05]. From the ABCD parameters, the corresponding Z matrix can be calculated as

$$Z = \frac{1}{C}\begin{pmatrix} A & AD - BC \\ 1 & D \end{pmatrix}. \tag{2.20}$$

2.3. Superconductivity

Superconductivity is the effect upon which most of this work is founded. In the following, I will therefore briefly summarize the aspects relevant to the understanding of the presented results. For further information, the reader is referred to textbooks such as [Sch97; Tin04; Lik91].

Superconductivity is a macroscopic quantum effect that occurs in certain materials below a critical temperature T_c and was first discovered in 1911 by Heike Kamerlingh Onnes [KO11] through the sudden disappearance of the electrical resistance of mercury. It took several decades until the first microscopic explanation was published in 1957 by Bardeen, Cooper and Schrieffer [BCS57]. It states that in some materials, when the thermal energy $k_\mathrm{B}T$ is sufficiently low, electrons can form a bound Cooper pair through a long range attraction mediated by phonons . The Cooper pairs condense into a bosonic ground state that can be described by a single macroscopic wave function

$$\Psi(\vec{r}) = \sqrt{n_s} e^{i\theta(\vec{r})}, \tag{2.21}$$

where n_s is the Cooper pair density and $\theta(\vec{r})$ the phase of the wave function.

2.3.1. Flux Quantization in a Superconducting Loop

One direct consequence of the single-valuedness of Eq. (2.21) is that the phase change acquired along a closed path $\vec{l}$ along a planar superconducting loop with normal $\vec{f}$ encompassing a non-superconducting hole must still be a multiple of 2π:

$$\oint \nabla\theta \, d\vec{l} = 2\pi n. \tag{2.22}$$

From the canonic momentum of a Cooper pair $\vec{p}$ and Eq. (2.21) one can derive the current $\vec{j}$ in the superconductor in the presence of a magnetic field

$$\vec{j} = \frac{2e\vec{p}}{2m_\mathrm{e}} = \frac{en_\mathrm{s}}{m_\mathrm{e}}\left(\hbar\nabla\theta - 2e\vec{A}\right), \tag{2.23}$$

where m_e is the electron mass and $\vec{A}$ is the vector potential. As currents only flow near the surface of a superconductor, Eq. (2.23) simplifies to $\nabla\theta = 2eA/\hbar$

deep inside the superconductor. When inserted into Eq. (2.22) we can see that the magnetic flux Φ through the enclosed area can only take quantized values:

$$
\begin{aligned}
2\pi n &= \frac{2e}{\hbar}\oint \vec{A}\,d\vec{l} = \frac{2e}{\hbar}\int (\nabla \times A)\,d\vec{f} \qquad (2.24)\\
&= \frac{2e}{\hbar}\int \vec{B}\,d\vec{f} = \frac{2e}{\hbar}\Phi,\\
\Phi &= n\frac{h}{2e} = n\Phi_0. \qquad (2.25)
\end{aligned}
$$

Here, $\Phi_0 = 2.068 \times 10^{-15}$ Vs is called the flux quantum.

2.3.2. The Josephson Junction

As was already stated in the introduction, a Josephson junction is a weak link between two superconductors. Since the experiments in this thesis only deal with *superconductor-isolator-superconductor* (SIS) tunnel-junctions, we will restrict the discussion to this special case in which the superconductors are separated by a few nanometer thick insulating barrier through which Cooper pairs and normal electrons can tunnel. The superconducting tunnel current and the voltage across the junction are related to the phase difference between the wave-functions on both sides of the junction and described by the Josephson relations Eq. (1.2a–1.2b).

The Josephson Inductance

A useful and widespread way of integrating Josephson junctions into circuit models is to find their small-signal behavior. We can find the Taylor series for Eq. (1.2a) around a phase value φ_0 only retaining the terms up to the first order:

$$I(t) = I_c \sin(\varphi_0) + I_c \cos(\varphi_0)\,(\varphi(t) - \varphi_0)\,. \qquad (2.26)$$

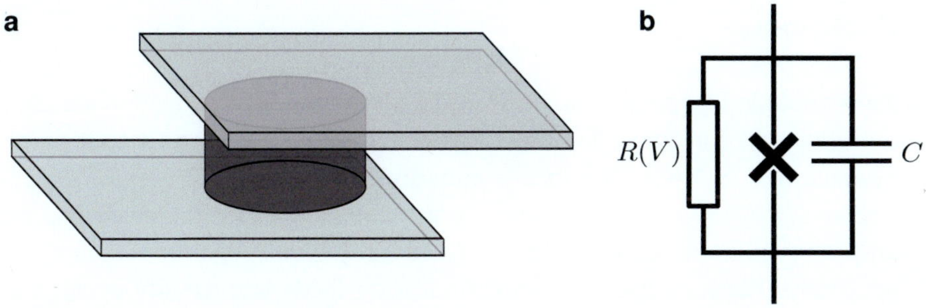

Figure 2.4.: **a**, The Josephson Tunnel junction (not to scale) is an isolating barrier (dark gray) sandwiched between two superconducting electrodes (light gray). Usually the junctions are only a few nanometers high while the electrodes have a film thickness of several hundred nanometers. **b**, RCSJ model of the Josephson junction. Parallel to the ideal junction are a capacitor and a resistor accounting for the capacitance between the electrodes and the quasi-particle tunnel current, respectively.

When rearranged for $\varphi(t)$ and inserted into Eq. (1.2b) we get

$$\begin{aligned} V(t) &= \frac{\Phi_0}{2\pi}\frac{\partial\varphi(t)}{\partial t} \\ &= \frac{\Phi_0}{2\pi I_\mathrm{c}\cos(\varphi_0)}\frac{\partial}{\partial t}I(t) \\ &= L_\mathrm{j}(\varphi_0)\frac{\partial}{\partial t}I(t). \end{aligned} \tag{2.27}$$

This yields the Josephson inductance introduced in Eq. (1.3). L_j is called an inductance because it relates the voltage over the junction to a time derivative of the current through it. Analogous to the kinetic inductance of the superconductor, however, the energy associated with the Josephson inductance is not stored in a magnetic field but as the kinetic energy of the tunneling Cooper pairs. What makes this effect really unique is the range of tunability: By changing φ_0 between 0 and π, one can tune L_j from its initial value $L_\mathrm{j}(0) = \Phi_0/(2\pi I_\mathrm{c})$ to infinity and from negative infinity to $-L_\mathrm{j}(0)$. The extreme values around $\varphi_0 = \pi/2$ are of course a consequence of the first-order cutoff performed in Eq. (2.26). In reality, the concept of the Josephson inductance does not hold for those values.

The RCSJ model

In order to include Josephson junctions into a circuit model, we need to account for more than just current and voltage associated with tunneling Cooper pairs. An approach that is often used in this situation is the *Resistively and Capacitively Shunted Junction (RCSJ) model.* In parallel to the element described by the Josephson relations is a capacitor and a resistor. The capacitor accounts for the capacitance between the two superconducting leads that usually overlap in case of tunnel junctions and thus form a parallel plate capacitor. The resistance accounts for the (lossy) contribution by tunneling quasi-particles. In general its value cannot be determined so easily because it depends on the number of available quasi-particles. Not only is this a temperature dependent quantity, it is also related to the voltage across the junction as quasi-particles can be created if the voltage is larger than the gap voltage $V_{\mathrm{gap}} = 2\Delta/e$. One way to account for this effect is by using a piecewise linear model

$$R(V) = \begin{cases} R_{\mathrm{sg}},\ V < V_{\mathrm{gap}} \\ R_{\mathrm{n}},\ V \geq V_{\mathrm{gap}} \end{cases}, \tag{2.28}$$

where R_{sg} is the sub-gap resistance, R_{n} the normal resistance and usually $R_{\mathrm{sg}} \gg R_{\mathrm{n}}$. Even though this model is already highly simplified, it leads to a self-consistent situation in the circuit model[5] due to which it cannot be solved in frequency domain. For this reason, it is often assumed that $R = R_{\mathrm{n}}$ which neglects the voltage dependence altogether.

[5] As R depends on V and vice versa.

2.4. The SQUID meta-atom

The basic building block of the superconducting metamaterial discussed in this work is the superconducting quantum interference device. SQUIDs are usually classified by the number of junctions they contain. Historically, single junction versions are called rf-SQUIDs while the ones containing two junctions are called dc-SQUIDs. Here will we focus on the former case, although most the concepts we are going to use can also be applied to multi-junction SQUIDs.

Unless stated otherwise, I will use a set of parameters for the SQUID (found column G2 in Tab. 3.2) wherever needed. They match those of the device that was used in all the single SQUID experiments presented in this thesis.

2.4.1. The Single Junction SQUID

In the context of this work, it is necessary to discuss the dynamics of a single junction SQUID (cf. Fig. 2.5a) that is subjected to an external magnetic flux Φ_{ext} generated by a magnetic field perpendicular to its loop area.

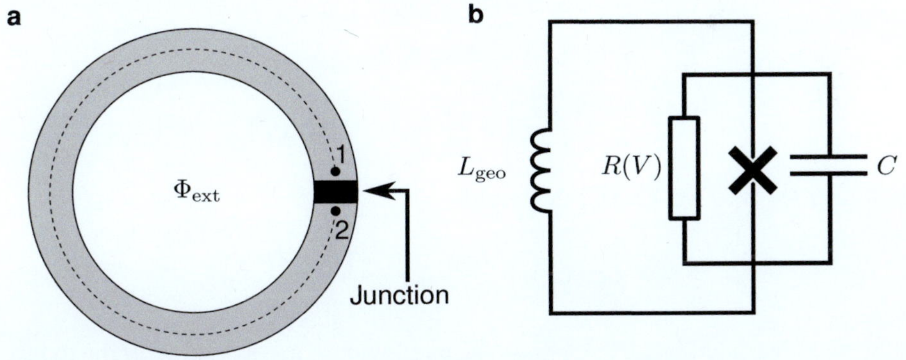

Figure 2.5.: **a**, Single junction (rf-)SQUID (based on the description in [Sch97]). A superconducting loop (light gray) interrupted by a Josephson junction (black) is threaded by a magnetic flux Φ_{ext}. Integration from point 1 to point 2 along the dashed path yields an expression for the phase difference across the junction (Eq. (2.29)) **b**, Equivalent circuit of an rf-SQUID. The loop is replaced by its geometric inductance L_{geo} and the junction by its RCSJ representation.

We can calculate the phase difference between both sides of the junction similar to Eq. (2.24) by integrating along the ring (on the path $1 \rightarrow 2$ as shown in Fig. 2.5a).

$$\varphi = \frac{2e}{\hbar} \int_{1}^{2} \vec{A}\, d\vec{l}. \tag{2.29}$$

Taking into account that the distance between points 1 and 2 is very small one can approximate the expression by a closed loop integral and write

$$\varphi = \frac{2e}{\hbar} \oint \vec{A}\, d\vec{l} = \frac{2\pi}{\Phi_0} \Phi, \tag{2.30}$$

where Φ is the total flux in the loop, which is the difference between the external flux and the flux generated by the screening current I_{sc} flowing in the loop:

$$\Phi_{\mathrm{ext}} = \Phi + L_{\mathrm{geo}} I_{\mathrm{sc}}. \tag{2.31}$$

The latter can be rewritten by applying Kirchoff's current law to the RCSJ representation of the junction seen in Fig. 2.5b. It is the sum of the currents flowing through the junction, the resistor and the capacitor. Together with Eq. (1.2b) we get

$$\begin{aligned} I_{\mathrm{sc}} &= C\dot{U} + \frac{1}{R} U + I_{\mathrm{c}} \sin(\varphi) \qquad (2.32) \\ &= \frac{\Phi_0 C}{2\pi} \ddot{\varphi} + \frac{\Phi_0}{2\pi R} \dot{\varphi} + I_{\mathrm{c}} \sin(\varphi). \end{aligned}$$

When inserted into Eq. (2.31), this yields

$$\varphi_{\mathrm{ext}} = \varphi + L_{\mathrm{geo}} C \ddot{\varphi} + \frac{L_{\mathrm{geo}}}{R} \dot{\varphi} + \frac{2\pi L_{\mathrm{geo}} I_{\mathrm{c}}}{\Phi_0} \sin(\varphi), \tag{2.33}$$

where we substituted $\varphi_{\mathrm{ext}} = 2\pi \Phi_{\mathrm{ext}} / \Phi_0$ as the phase associated with the external flux Φ_{ext}. Eq. (2.33) is more commonly written in a different form:

$$\varphi_{\mathrm{ext}} = \varphi + \beta_{\mathrm{L}} \left[\sin\varphi + \frac{1}{\omega_{\mathrm{c}}} \dot{\varphi} + \frac{1}{\omega_{\mathrm{p}}^2} \ddot{\varphi} \right], \tag{2.34}$$

where $\beta_{\mathrm{L}} = 2\pi L_{\mathrm{geo}} I_{\mathrm{c}} / \Phi_0$ is the SQUID parameter, $\omega_{\mathrm{p}} = \sqrt{2\pi I_{\mathrm{c}} / (C\Phi_0)}$ is the plasma frequency and $\omega_{\mathrm{c}} = 2\pi R I_{\mathrm{c}} / \Phi_0$ the characteristic frequency related to the damping. Another common way of expressing the characteristic damping is the McCumber parameter $\beta_c = \omega_c \cdot RC$.

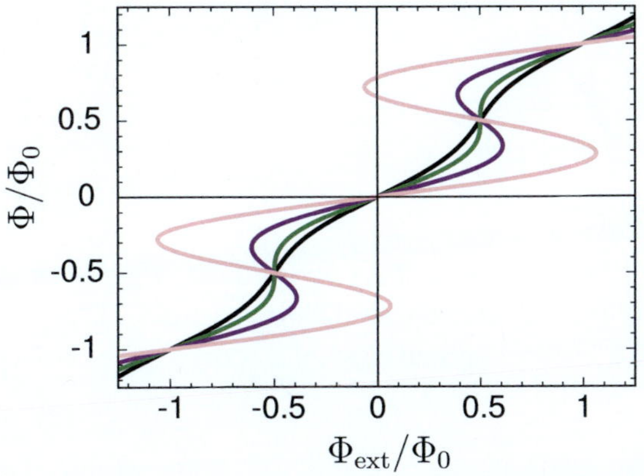

Figure 2.6.: Total flux in the loop vs. externally applied flux. The colors correspond to different values of the SQUID parameter: $\beta_L = 0.5$ (black), $\beta_L = 1$ (green), $\beta_L = 2$ (purple), $\beta_L = 5$ (pink).

The parameter β_L can also be written as the ratio between the geometric inductance and the Josephson inductance at zero phase difference

$$\beta_L = \frac{L_{geo}}{L_j(0)}, \tag{2.35}$$

and serves an important role in the characterization of rf-SQUIDs: In the static case, the functional dependence $\varphi(\varphi_{ext})$ is defined through the static version of Eq. (2.34)

$$\varphi_{ext} = \varphi + \beta_L \sin\varphi, \tag{2.36}$$

which is plotted in Fig. 2.6. For values of $\beta_L \leq 1$, it is single-valued, for larger values $\beta_L > 1$, it starts to become multi-valued around $\varphi_{ext} = \pm n\pi$ and the SQUID is called hysteretic. For even larger values $\beta_L \gtrsim 4.61$ it becomes multivalued in the whole φ_{ext} range.

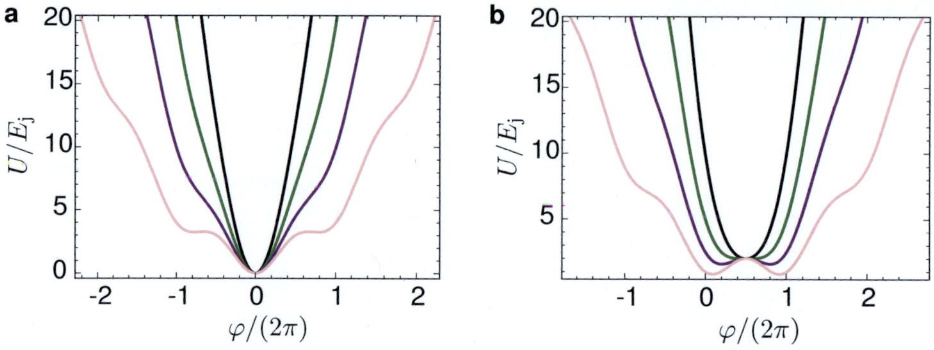

Figure 2.7.: SQUID Potential for **a**, $\Phi_{\text{ext}} = 0$ and **b**, $\Phi_{\text{ext}} = \Phi_0/2$. As in the previous figure, the colors correspond to different values of the SQUID parameter: $\beta_L = 0.5$ (black), $\beta_L = 1$ (green), $\beta_L = 2$ (purple), $\beta_L = 5$ (pink).

For the following discussion of the transient dynamics of the SQUID, it is instructive to treat the phase evolution in terms of the motion of a virtual phase particle with kinetic and potential energy.

$$E_{\text{kin}} = \frac{1}{2}\frac{E_j}{\omega_p^2}\dot{\varphi}^2 \tag{2.37}$$

$$U = E_j\left[1 - \cos\varphi + \frac{(\varphi - \varphi_{\text{ext}})^2}{2\beta_L}\right], \tag{2.38}$$

where $E_j = I_c\Phi_0/(2\pi)$ is the so called Josephson energy. Fig. 2.7 shows a plot of Eq. (2.38) for two different values of the external flux. One can see that the multi-valuedness of Eq. (2.36) leads to the formation of stable wells in the potential that can trap the phase particle. Although this effect may have its application in a SQUID meta-material (as was suggested in [DCL06]), we are interested in effects that appear in non-hysteretic SQUIDs. Thus, from now on, we will limit our discussion to the non-hysteretic case.

2.4.2. Limits of weak and strong driving

This subsections is dedicated to the analysis of the SQUID dynamics in the limits of very weak and very strong driving. We will see that in both cases, the dynamics reduce to that of a linear system with different properties. This is a unique feature of the Josephson nonlinearity and can qualitatively be understood from Eq. (2.34): If the driving (and thus the variation in φ) is very small, the term $\varphi + \beta_\mathrm{L} \sin\varphi$ can be linearized around its static value. In the opposite case (strong driving or large variation in $\varphi \gg \beta_\mathrm{L}$), the $\beta_\mathrm{L} \sin\varphi$ term can be treated as a small perturbation to φ or neglected altogether.

Response in the Limit of Weak Driving

In case of very small oscillations of the phase φ with amplitude φ_a around a constant value φ_0

$$\varphi = \varphi_0 + \varphi_\mathrm{a} \cos\omega t, \tag{2.39}$$

Eq. (2.33) can be treated in the same fashion that was used to derive the Josephson inductance in Eq. (2.27). The dynamic part reduces to a tunable harmonic oscillator

$$\frac{\varphi_\mathrm{ext}}{L_\mathrm{geo}C} = \ddot{\varphi} + \frac{1}{RC}\dot{\varphi} + \frac{1 + \beta_\mathrm{L}\cos\varphi_0}{L_\mathrm{geo}C}\varphi. \tag{2.40}$$

In this form, the prefactor of the last term is the square of the angular resonance frequency ω_r^2 and the resonance frequency, which is tunable by φ_0, is (cf. Fig. 2.8a)

$$\begin{aligned} f_\mathrm{r} &= \frac{\omega_\mathrm{r}}{2\pi} = \frac{1}{2\pi} \cdot \sqrt{\frac{1 + \beta_\mathrm{L}\cos\varphi_0}{L_\mathrm{geo}C}} \\ &= f_\mathrm{r0}\sqrt{1 + \beta_\mathrm{L}\cos\varphi_0} = \omega_\mathrm{p}\sqrt{\frac{1}{\beta_\mathrm{L}} + \cos\varphi_0}, \end{aligned} \tag{2.41}$$

where $f_\mathrm{r0} = (2\pi\sqrt{LC})^{-1}$ is the resonance frequency of the RLC circuit without the junction. The same result can be obtained by replacing the junction in Fig. 2.5 with the value of its Josephson inductance (Eq. (1.3)). Then the two parallel inductors with values L_geo and L_j can be treated as one with a combined inductance using Eq. (2.35):

$$L_\mathrm{c} = L_\mathrm{geo} || L_\mathrm{j} = \frac{L_\mathrm{j}L_\mathrm{geo}}{L_\mathrm{j} + L_\mathrm{geo}} = \frac{L_\mathrm{geo}}{1 + \beta_\mathrm{L}\cos\varphi_0}. \tag{2.42}$$

The dependence of this quantity on external flux is shown in Fig. 2.8b.

In this parallel RLC circuit, the resonance frequency is $f_\mathrm{r} = (2\pi\sqrt{L_\mathrm{c}C})^{-1}$, which leads to the same result as Eq. (2.41). From these equations we can deduce that the upper and lower limit of the resonance frequency are given by $f_\mathrm{max} = f_\mathrm{r0}\sqrt{1+\beta_\mathrm{L}}$ and $f_\mathrm{min} = f_\mathrm{r0}\sqrt{1-\beta_\mathrm{L}}$, respectively. The quality of the resonance is given by the simple relation

$$Q = \frac{\omega_\mathrm{c}\omega_\mathrm{r}}{\omega_\mathrm{p}^2} = \frac{\omega_\mathrm{c}}{\omega_\mathrm{p}}\sqrt{\frac{1}{\beta_\mathrm{L}} + \cos\varphi_0}\,. \tag{2.43}$$

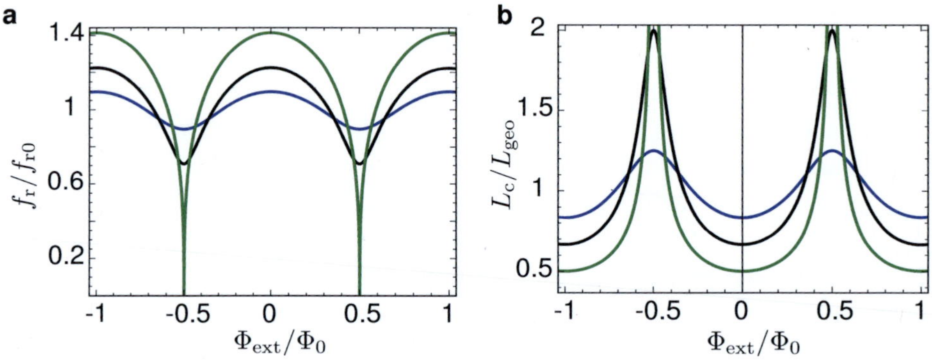

Figure 2.8.: Dependence of **a**, the resonance frequency and **b**, the total inductance of a rf-SQUID on external magnetic flux. The colors correspond to different values of the SQUID parameter: $\beta_\mathrm{L} = 0.2$ (blue), $\beta_\mathrm{L} = 0.5$ (black), $\beta_\mathrm{L} = 1$ (green).

It should be stressed that, although the small signal approximation of the Josephson nonlinearity breaks down as φ_0 approaches[6] $(n-0.5)\pi$, $n \in \mathbb{Z}$, the predictions for the rf-SQUID (as seen in Fig. 2.8) still work reasonably well as long as β_L significantly less than 1. The reason for this is that for a large Josephson inductance for this range of φ_0, the combined inductance L_c is dominated by the (much smaller) geometric inductance.

[6]Which is where the dc current in the junction approaches its critical value and the Josephson inductance would diverge.

Response in the Limit of Strong Driving

In case of very large variations of $\varphi \gg \beta_L$, the $\beta_L \sin\varphi$ term in Eq. (2.34) can be neglected as it oscillates in φ with an amplitude β_L while the linear term φ increases with the driving strength. Consequently, the dynamic part of the equation of motion

$$\frac{\varphi_{\text{ext}}}{L_{\text{geo}}C} = \ddot{\varphi} + \frac{1}{RC}\dot{\varphi} + \frac{1}{L_{\text{geo}}C}\varphi \tag{2.44}$$

does not depend on φ_0 any more. Instead, it reduces to that of the pure RLC circuit that resonates at the LC resonance frequency f_{r0}.

2.4.3. Intermediate Driving

In the last two subsections, I described the quasi-linear dynamics of the non-hysteretic rf-SQUID in the limit of weak and strong driving. The transition between those two regimes, however, cannot be dealt with so easily. It requires a more rigorous treatment that will be introduced as part of the results of this thesis in section 4.1.1. Here, I will introduce the concept of the Duffing oscillator as an example of how the nonlinearity can be accounted for. Although it does not work particularly well for the rf-SQUID, it still exhibits some similar phenomena and can be used to describe the onset of bistability.

From here on, we will make an assumption that may at first seem unintuitive for a nonlinear system such as the SQUID. When a nonlinear oscillator is driven with a frequency ω, oscillations with integer multiples of the base frequency are induced in addition to the fundamental frequency:

$$\varphi(t) = \sum_{n=1}^{\infty} \varphi_a^{(n)} \cos\left(n\omega t + \delta^{(n)}\right). \tag{2.45}$$

For the following discussion, however, we will assume that the oscillations of higher harmonics are much smaller than the fundamental one and thus

$$\varphi(t) \simeq \varphi_a^{(1)} \cos\left(\omega t + \delta^{(1)}\right) \equiv \varphi_a \cos\left(\omega t + \delta\right). \tag{2.46}$$

This assumption will be verified later in section 4.1.

The Duffing-Oscillator

The Duffing oscillator is an extension to the harmonic oscillator that includes a x^3 term in the equation of motion. This can be used to extend the low-power case of the SQUID to higher powers under a harmonic drive $\varphi_{\mathrm{ext}} = \varphi_{\mathrm{ea}} \cos(\omega t)$. To apply it, we have to replace the sinusoidal term in Eq. (2.34) by the first two terms of its series expansion

$$\sin\varphi = \varphi - \frac{1}{3!}\varphi^3 + o(\varphi^5) \approx \varphi - \frac{1}{6}\varphi^3. \tag{2.47}$$

This last approximation is only valid for small values of φ (such as $\varphi < \pi/4$) but most qualitative results can be reproduced for φ as large as $\pi/2$. We will therefore assume an external flux as in Eq. (2.39) but without any time-independent offset ($\varphi_0 = 0$). Replacing the sine term in Eq. (2.34) by Eq. (2.47), we can apply the well-known formulas (e.g. as can be found in [LL98]) and calculate the amplitude of the oscillations in the SQUID. They are given as the real solution of a cubic equation in φ_a^2:

$$\varphi_a^2 \left[\left(\delta_\omega + \frac{\omega_{\mathrm{p}}}{16\sqrt{1+\beta_{\mathrm{L}}^{-1}}}\varphi_a^2 \right)^2 + \frac{\omega_{\mathrm{p}}^4}{4\omega_{\mathrm{c}}^2} \right] = \frac{\omega_{\mathrm{p}}^2 \varphi_{\mathrm{ea}}^2}{4\left(\beta_{\mathrm{L}}^2 + \beta_{\mathrm{L}}\right)}. \tag{2.48}$$

Here, $\delta_\omega = \omega - \omega_{\mathrm{r}}$ is the difference between the frequency of the drive and the low power resonance frequency and has to remain small for the model to yield correct results.

Depending on the value of the driving amplitude, the oscillations in φ show qualitative differences in their behavior as is shown in Fig. 2.9. For very low values of the driving amplitude φ_{ea}, the influence of the nonlinearity is small and the SQUID behaves like a harmonic oscillator, as expected in the weak-driving limit. (c.f. blue line in Fig. 2.9). As the amplitude of the drive increases, the resonance curve starts to lean to the left (c.f. green line in Fig. 2.9) until φ_{ea} reaches a critical value

$$\varphi_{\mathrm{crit}} = 8\beta_{\mathrm{L}} \sqrt{\frac{\omega_{\mathrm{p}}^3}{\omega_{\mathrm{c}}^3} \frac{\left(1+\beta_{\mathrm{L}}^{-1}\right)^{3/2}}{3\sqrt{3}}} \tag{2.49}$$

above which more than one stable solution of Eq. (2.48) exists for some frequencies below the original resonance.

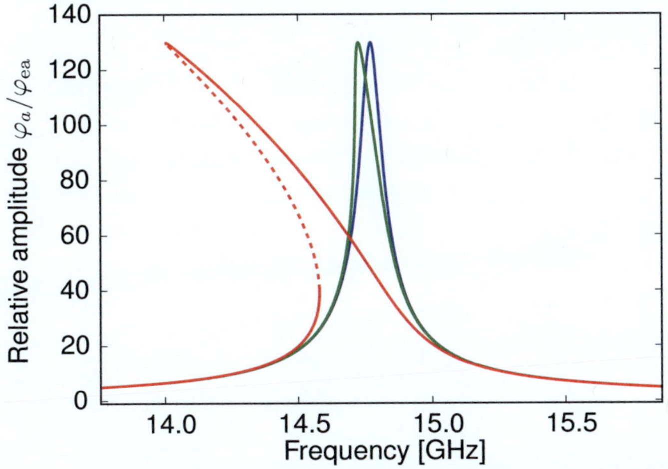

Figure 2.9.: Relative amplitude of the oscillations $\varphi/\varphi_{\text{ea}}$ in the SQUID for different driving strength levels around the low power, zero bias resonance frequency. The blue line corresponds to $\varphi_{\text{ea}} = 2\pi \cdot 10^{-4}$, the green one to $\varphi_{\text{ea}} = 10\pi \cdot 10^{-4}$ and the red one to $\varphi_{\text{ea}} = 4\pi \cdot 10^{-3}$. The dashed section, although solving Eq. (2.48), is unstable.

In this case (c.f. red line in Fig. 2.9) two of the three solutions of Eq. (2.48) are stable (solid lines) while the third is unstable (dashed line) and cannot be reached. Which of the possible solutions is realized depends on the history and initial conditions. If, for example, we were to start driving the system at a frequency below the bistable region and increase the frequency, the amplitude of the oscillations would follow the red curve from the left until it reaches the dotted section. From there, it would jump to the upper branch and follow it further to the right. When sweeping the frequency down, the system would stay in the upper branch for as long as possible, jumping back to the lower branch at a different frequency. This hysteresis is one of the measurable indicators of bistability.

It should be noted that the tip of the red curve in Fig. 2.9 is already at a point ($\varphi_a \approx \pi/2$) for which the approximation of Eq. (2.47) starts to fail. Thus, in order to investigate the SQUID dynamics under even stronger driving, a different description is necessary.

At this point, it should also be stressed that the bistability and hysteresis we just discussed have nothing to do with that introduced earlier in SQUIDs with $\beta_\mathrm{L} > 1$ (cf. Fig. 2.7). Although even in the driven case, both can lead to a hysteresis in the amplitude of the oscillations, the *dynamic multi-stability* also works for smaller values of β_L (e.g. $\beta_\mathrm{L} \approx 0.45$ was used in Fig. 2.9) where there is only one stable state in the stationary SQUID potential.

3. Experimental Setup & Procedures

"How hard can it be?"

Jeremy Clarkson, TV host on the show *Top Gear*

(Usually just as something is about to go terribly wrong.)

The path from an idea to the experimental results inevitably includes considerations about the experimental setup, which are the subject of chapter 3. This includes the samples, the hardware used in the measurements and the experimental techniques. Many aspects of those things may seem obvious or even trivial to anyone working in the same field.
Nonetheless, a very clear understanding of the experiment is necessary to relate measured results to theoretical quantities.

3.1. Samples

In this section, I will introduce the different rf-SQUID samples used throughout this thesis. In total, there are three different generations of samples. Although all of them are similar in many respects, some key changes have been implemented between the generations which significantly influence the results.

In order to implement an rf-SQUID (Fig. 2.5a) using a Josephson tunnel junction in a niobium (Nb) thin film process, a number of things have to be considered. As the Josephson junction consists of an isolating barrier between two superconducting layers (cf. Fig. 2.4a), one cannot just form a loop between both sides of the junction. Instead, the loop is made from two parts in different layers. An interconnecting part (also called a *via*) forms a superconducting connection between the upper and the lower layer and closes the loop. Also, the thickness of the films that make up the leads forming the loop is much smaller than all the lateral dimensions of the structure. This circumstance facilitates the trapping of Abrikosov vortices [But14] in the structure, which has to be avoided where possible.

One of the major design considerations for this experiment was the SQUID's resonance frequency under weak driving (cf. Eq. (2.41)), which is defined through the critical current of the junction I_c, the loop inductance L and the capacitance C. As we need a non-hysteretic SQUID ($\beta_L < 1$), the loop inductance has to be smaller than the Josephson inductance at zero flux ($L_j(0) = \Phi_0/(2\pi I_c)$, cf. Eq. (2.35)). This imposes an upper limit on L, as I_c is limited from below. This technological limitation has two reasons: First, the junction area cannot be smaller than a minimum value A_{min} due to the resolution of the photo-lithographic process which is typically of the order of 1 μm^2. Second, the critical current density j_c of the tunnel junction is limited by the thickness of the tunnel barrier [KMM95]. Due to the self-saturating nature of the oxidization process, this thickness is limited to a maximum value for a given fabrication procedure. The available critical current densities were 1 kA/cm^2 for the first generation of samples and 0.1 kA/cm^2 for the second and third generation.

Using only the specific capacitance C_s of the junction (typically $\approx$ 50fF/μm^2 [Mae+95]), the required frequency range (below 20 GHz) cannot be reached for rf-SQUIDs in this process. Due to the fixed ratio between C_s and j_c for a given fabrication process, the resonance frequency of the SQUID can be written as a monotonically decreasing function of β_{L}:

$$2\pi f_{\mathrm{max}} = \omega_{\mathrm{max}} = \frac{1}{2\pi\sqrt{\frac{L \cdot L_{\mathrm{j}}(0)}{L+L_{\mathrm{j}}(0)}C}} = \frac{1}{2\pi\sqrt{\frac{\beta_{\mathrm{L}}}{1+\beta_{\mathrm{L}}}\frac{\Phi_0 C_{\mathrm{s}}}{2\pi j_{\mathrm{c}}}}}. \tag{3.1}$$

In the limit of a junction dominated SQUID $\beta_{\mathrm{L}} \to \infty$, this value equals the junction plasma frequency

$$\lim_{\beta_{\mathrm{L}} \to \infty} \omega_{\mathrm{max}} = \sqrt{\frac{2\pi j_{\mathrm{c}}}{C_{\mathrm{s}}\Phi_0}} = \omega_{\mathrm{p}}, \tag{3.2}$$

while for all other values $\omega_{\mathrm{max}} > \omega_{\mathrm{p}}$. This implies that for the typical values of j_{c} and C_{s} given above, the resonance frequency is larger than $\omega_{\mathrm{p}}/2\pi \approx 100\,\mathrm{GHz}$ for any value of β_{L}. To circumvent this limitation, a shunt capacitance C_{shunt} was added to the SQUID design parallel to the junction. It was implemented in form of a parallel plate capacitor between the top and bottom Nb layers with anodized Nb (such as Nb_2O_5) as dielectric. Due to its high relative permittivity ($\varepsilon_{\mathrm{r}} \approx 32$) and the small thickness of the layer (30-40 nm), this material is ideally suited to build small area capacitors with high capacitance values ($\approx 2\,\mathrm{pF}$ in our case).

In addition to the SQUIDs, some of our samples also contain coplanar waveguide structures which are created in the same process. For all samples, the central conductor is made from Nb as it is used to bias the SQUIDs with magnetic field. The ground planes of the CPW, however, are only superconducting in the first generation samples.

3.1.1. Fabrication

The first generation of samples was produced by the U.S. company Hypres Inc[1], the second and third at the Kotel'nikov Institute of Radio Engineering and Electronics (IREE RAS) in Moscow, Russia. Both foundries use similar Nb processes with a Nb/AlO_x/Nb trilayer for the Josephson junctions.

First, the trilayer is deposited onto an oxidized Si wafer. Then, the junction area is defined. In the Hypres process, the next step is the anodization of the

[1] http://www.hypres.com

shunt capacitor area before the definition of the base electrode. In the IREE process, however, these steps are processed in reverse order. This makes it necessary to leave galvanic bridges between all parts that have to be anodized at this stage and connect them to a lead on the edge of the waver. They are removed later in the process but their shadows are still visible under the microscope. Next, a SiO_2 layer separates the next Nb layer from the base electrode wherever contact is undesirable. Finally, a Au layer is deposited at the end of the CPWs to facilitate wire-bonding to the structure.

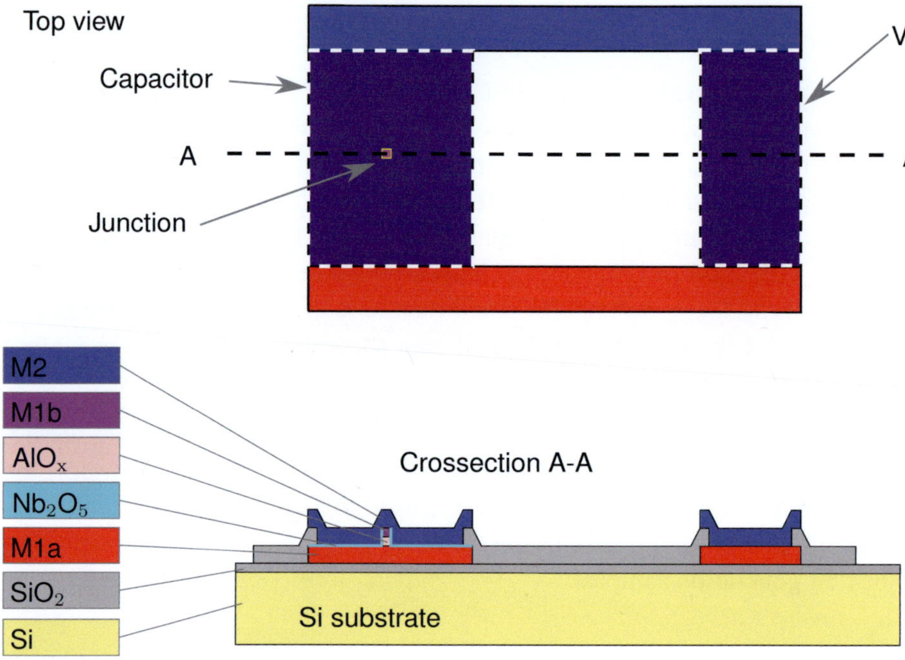

Figure 3.1.: Sketch of the layer structure of the first generation SQUID featuring a top view and a cut along the A-A direction. On top of the SiO_2 buffered Si substrate is the trilayer consisting of M1a, the AlO_x tunnel junction and M1b. The shape of the latter two also define the junction area. The base electrode is anodized (Nb_2O_5) where the shunt capacitance should be. Another SiO_2 layer separates the trilayer from the top Nb layer M2 everywhere except for the capacitor, the junction and via area.

A sketch of how the layers look in the finished first generation SQUID is shown in Fig. 3.1.

After these steps, the wafers are diced into chips that are $5 \times 5\,\mathrm{mm}^2$ in the Hypres process and $4 \times 4\,\mathrm{mm}^2$ in the IREE process. In the first generation, each chip contained three CPW structures each in turn containing different types and numbers of SQUIDs. In later generations, the number of CPWs per chip was reduced to two.

3.1.2. Sample Generations

In this subsection, I will briefly summarize the key features of the SQUID generations which I will label G1 – G3. At the end of the section, all relevant numbers will be summarized in a table.

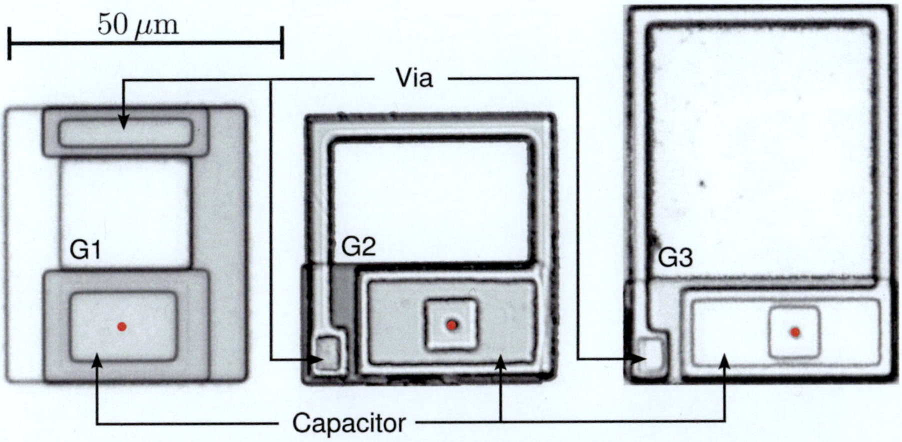

Figure 3.2.: Comparison of the three SQUID generations. From left to right: First generation Hypres SQUID, second generation IREE SQUID, third generation IREE SQUID. All three micro graphs were optimized to increase visibility. The arrows indicate the positions of the vias and capacitors. The positions of the Josephson junctions are highlighted by red dots.

The most distinctive feature about the first generation SQUID (G1) is that the electrodes in M1a and M2 that form the loop are almost identical in size and shape (cf. Fig. 3.2). As a result, the junction and via are on the opposite sides

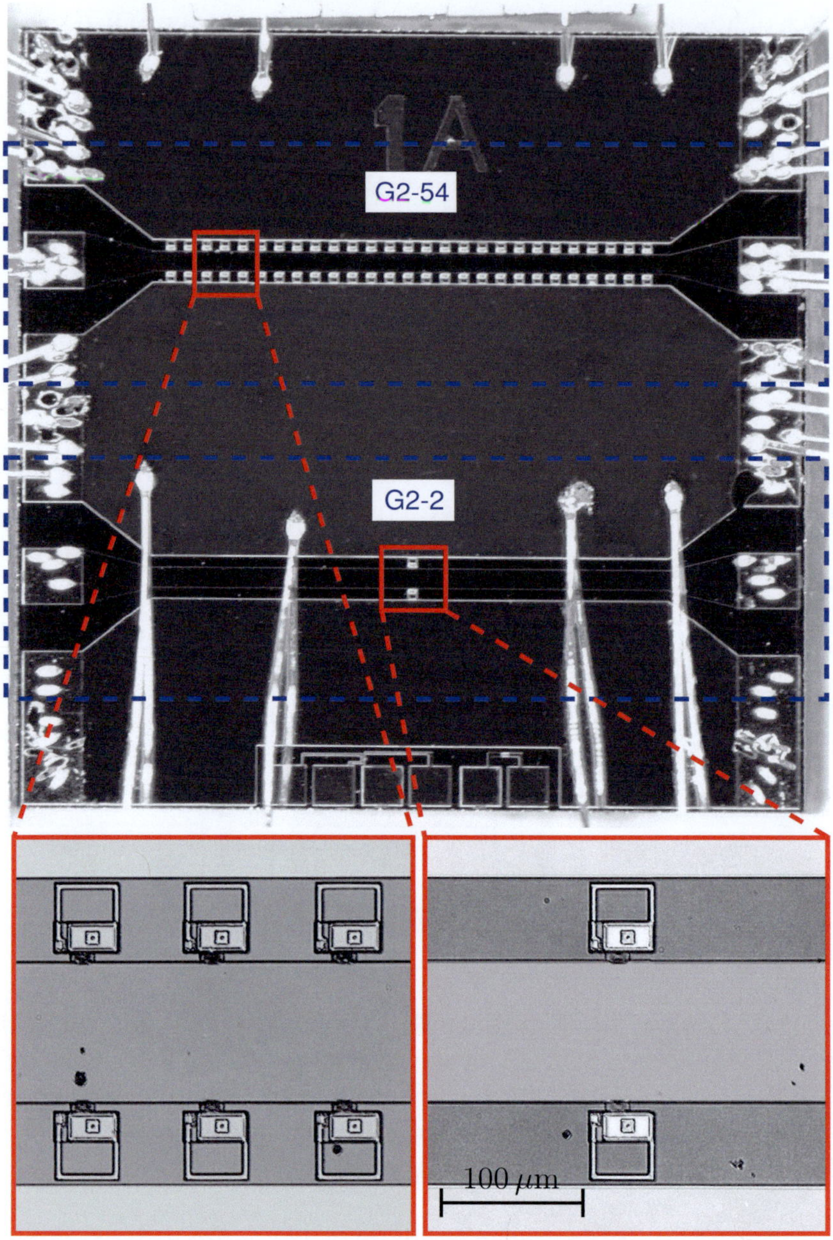

Figure 3.3.: Complete $4 \times 4\,\mathrm{mm}^2$ chip containing samples G2-54 (upper CPW) and G2-2 (lower CPW). The wire bonds are placed for measurements of sample G2-54. Below are two higher resolution details of the CPW gaps containing the SQUIDs.

of the loop. Both the via area and the width of the leads were much larger than in later samples. I will show results from two samples, the first one is a CPW containing one SQUID (G1-1), the second one a CPW containing 60 SQUIDs (G1-60).
The second generation SQUID (G2) has much narrower leads than its predecessor and is made almost completely from the base trilayer base electrode. M2 is only used as the upper plate of the capacitor and connects to the via located directly beside it. This design was also chosen because the new loop structure has a higher loop inductance[2] and thus the SQUID a lower band of resonance frequencies. Again, there are two samples. The first with a CPW containing two SQUIDs (G2-2), the second one containing 54 SQUIDs (G2-54). At some point, it became necessary to measure a CPW with only a single SQUID, so one of the two in G2-2 was removed using a scalpel. For measurements on the sample after this operation I will refer to the sample as G2-1. Fig. 3.3 features microscope pictures of a chip containing both G2 samples. It shows the chip that is wire-bonded to a PCB (cf. Fig. 3.7a) for a measurement of the upper of the two samples.

Apart from the lower resonance frequencies, one design consideration for the second generation of SQUIDs was to suppress the trapping of Abrikosov vortices in the structures. Since these changes proved to be successful (as I will show later in section 4.2.2), the third generation (G3) SQUIDs are very similar to their predecessors. They only differ in the exact dimensions of the capacitor and the loop. Although there are also samples with 1D arrays of SQUIDs from this generation, the only G3 samples used in this thesis are two-dimensional arrays. There are two types of arrays, both of them containing $21 \times 21 = 441$ G3 meta-atoms. Both are designed such that the coupling between neighboring SQUIDs in x- and y-direction is approximately identical. In one of them, the SQUIDs are densely packed (G3-D, cf. Fig. 3.4a). The coupling between nearest neighbors is $k_{\mathrm{nn}} = -2.9 \cdot 10^{-2}$ and $k_{\mathrm{dn}} = -6.87 \cdot 10^{-3}$ to the diagonal neighbors. In the other sample, the spacing is larger (G3-L, cf. Fig. 3.4b). Here, the coupling between nearest neighbors is $k_{\mathrm{nn}} = -6.4 \cdot 10^{-3}$ and $k_{\mathrm{dn}} = -1.87 \cdot 10^{-3}$ to the diagonal neighbors.

[2]Increasing the loop inductance was possible because the IREE process allows lower critical current values. This shifts the limit imposed by the requirement to keep β_{L} below one.

[3]Calculated for the effective junction (i.e. including the shunt capacitance).

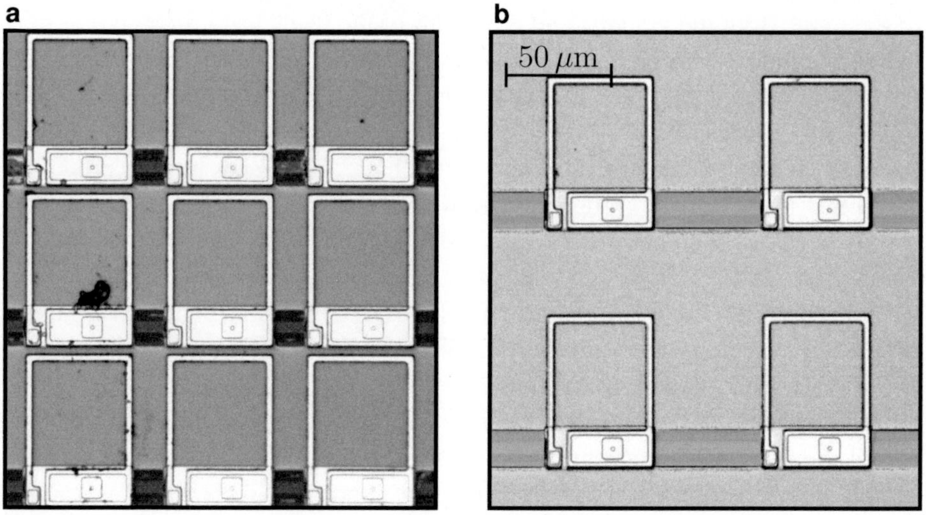

Figure 3.4.: Optical micrographs of third generation SQUID samples (2D arrays). **a**, Densely packed, strongly coupled SQUIDs (G3-D). **b**, Loosely packed, weakly coupled SQUIDs (G3-L).The horizontal bars between the SQUIDs are visible leftovers from the M1a bridges used for anodization.

	Description	G1	G2	G3
w	SQUID width	$45\,\mu$m	$45\,\mu$m	$49\,\mu$m
h	SQUID height	$52\,\mu$m	$49\,\mu$m	$71\,\mu$m
L	Loop inductance	63.5 pH	83 pH	131 pH
C	Shunt capacitance	1.7 pF	2.03 pF	2.05 pF
R	(RCSJ-)Resistance	$0.5\,\mathrm{k}\Omega$	$1\,\mathrm{k}\Omega$	$1\,\mathrm{k}\Omega$
I_c	Critical current	$4.5\,\mu$A	$1.79\,\mu$A	$1.86\,\mu$A

Table 3.1.: Physical parameters of the three SQUID generations. The width and height values correspond to the outer dimensions of the structure as seen in Fig. 3.2. The R values are just rough approximations based on the the normal state resistance of the junction R_n.

The physical parameter values for the different SQUID generations are given in Tab. 3.1, the resulting SQUID values in Tab. 3.2.

	Description	G1	G2	G3
β_{L}	SQUID parameter	0.87	0.45	0.74
β_{c}	McCumber parameter[3]	5805	11030	11574
$\omega_{\mathrm{c}}/(2\pi)$	Characteristic freq. [GHz]	1086.96	864.73	898.55
$\omega_{\mathrm{p}}/(2\pi)$	Plasma freq.[3] [GHz]	14.27	8.23	8.35
f_{r0}	LC resonance freq. [GHz]	15.32	12.26	9.71
f_{max}	Max. resonance freq. [GHz]	20.94	14.76	12.81
f_{min}	Min. resonance freq. [GHz]	5.52	9.09	4.97

Table 3.2.: Characteristics of the three SQUID generations calculated from the parameters given in Tab. 3.1.

3.2. Microwave setup

In order to measure the transmission through the SQUID-loaded CPWs that were introduced in the last subsection, they have to be integrated into a suitable setup.

Starting from a chip such as the one shown in Fig. 3.3, the first task is to organize the connection to coaxial cables. This is necessary because most components in the frequency range of interest use coaxial cables with standardized connectors as waveguides.
For this purpose, the chip is inserted into a matching, milled out space in a circular printed circuit board (cf. Fig. 3.5). The PCB itself contains coplanar waveguides which are wire-bonded to the chip on one side and have a coplanar to coaxial (SMP) connector on the other end.

Figure 3.5.: Sample holder PCB containing a $4 \times 4\,\mathrm{mm}^2$ chip. The chip is wire-bonded to the PCB (shown in more detail in the inset). The two unused coplanar waveguides on the PCB that run perpendicular to the others are connected to ground. On the far ends of the waveguides on the PCB, surface mounted (SMP) connectors form the transition from coplanar to coaxial waveguides.

The substrate of the PCB is a ceramic[4] with a dielectric constant close to that of silicon. This way, the CPWs on the PCB can be made almost as small as those on the chip which facilitates matching and bonding. Additionally, a low thermal

[4]Rogers TMM10i. http://www.rogerscorp.com

coefficient of the dielectric constant makes this type of material better suitable for low-temperature applications than most other dielectrics.

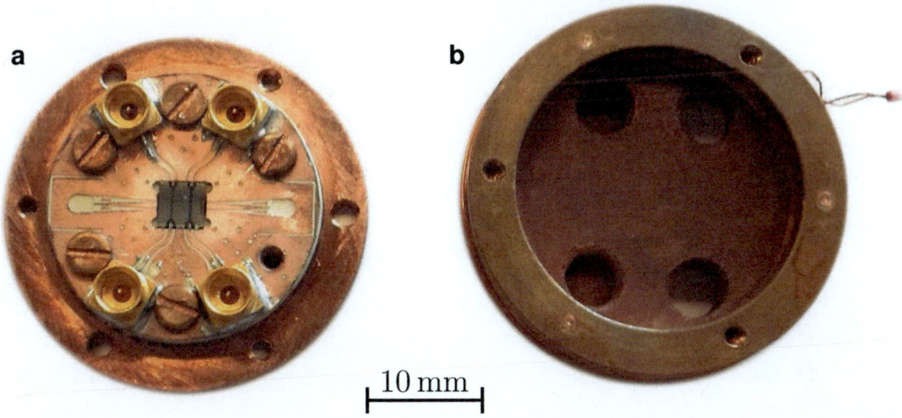

Figure 3.6.: Picture of the sample holder box. **a**, The bottom part holds the PCB. **b**, The top part has holes through which coaxial cables connect to the connectors on the PCB. Only two of the four holes are open. The two wires, visible in the upper right, connect to a coil wound around the outside of the top part that can be used to apply a magnetic field to the sample.

The PCB is then mounted on the bottom part of a sample holder box shown in Fig. 3.7b. When both parts of the sample holder box are connected, they form a closed cavity which is only accessible from the outside through the SMP connectors. The top part of the sample holder also has a recess around its circumference (cf. Fig. 3.7) which houses a coil, made from 500 windings of NbTi. It is used to apply a magnetic field to the sample. When both parts of the sample holder are connected together, the chip is positioned in the center of the coil where the field is most homogeneous.

Design drawings of the PCB without chip and connectors as well as the sample holder box are depicted in Fig. 3.7. The sample holder is connected to two microwave cables and mounted at the very bottom of a cryogenic dipstick in liquid ^{4}He inside a magnetic shield. The shield is made from Cryoperm, a type of μ-metal optimized for cryogenic applications. Its purpose is to shield the sample from any static magnetic fields, such as the earth's magnetic field and the fields emanating from ferromagnetic components in the setup.

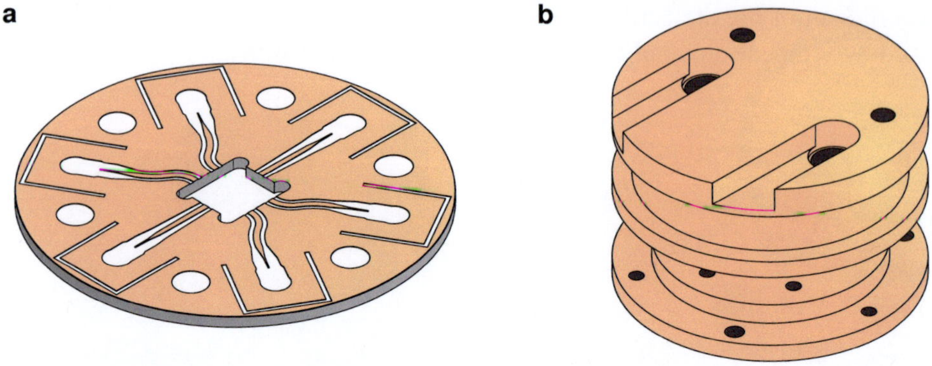

Figure 3.7.: Design drawings of **a**, PCB and **b**, Sample holder box. The latter consists of two parts. A bottom plate, onto which the PCB is mounted, and a top part, which holds the microwave cables. When connected together, the interior is a closed cavity.

The rest of the low temperature setup is depicted in Fig. 3.8. The microwave signal enters through a coaxial cable from the room-temperature electronics, passes a 30 dB attenuator before going into the rf port of a bias-T. It then passes through the sample and into the dc+rf port of a second bias-T, through a 3 dB attenuator and then into a cryogenic *high electron mobility transistor (HEMT)* amplifier before going back to room-temperature.

The purpose of the cold 30 dB attenuator is to reduce the thermal noise coming to the sample from the microwave source. The second (3dB) attenuator is used to reduce standing waves due to reflections at the input of the amplifier. The two bias-Ts serve a dual purpose. Their primary function is the possibility to superimpose the microwave in the sample with a dc current that runs through the central conductor of the CPW on the chip. This way, it creates a dc magnetic field that is anti-symmetric in the two CPW gaps with respect to the central conductor. Using this together with the field created by the coil, which is symmetric, means that the SQUIDs in one gap can be biased independently from the ones in the other gap. The second purpose of the bias-Ts is that they act as a high-pass filter on the microwave line, blocking low-frequency noise below their cutoff frequency ($\approx 40\,\text{kHz}$).

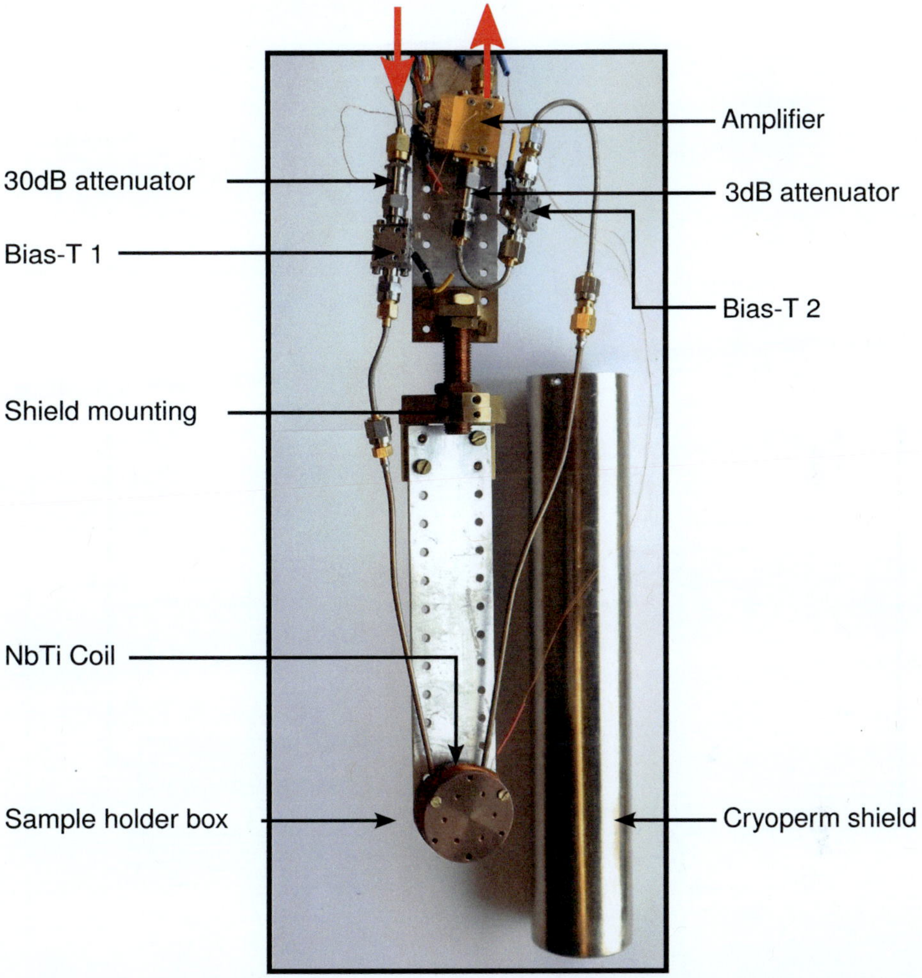

Figure 3.8.: Low temperature part of the experimental setup. The red arrows show the direction in which the microwave propagates through the devices. The Cryoperm shield shown on the right is fixed by three screws at the position labeled “Shield mounting”.

The room temperature part of the setup consists of a VNA that measures the transmitted microwave signal and two current sources that supply the bias-Ts and the coil. A sketch of the complete setup is shown in Fig. 3.9.

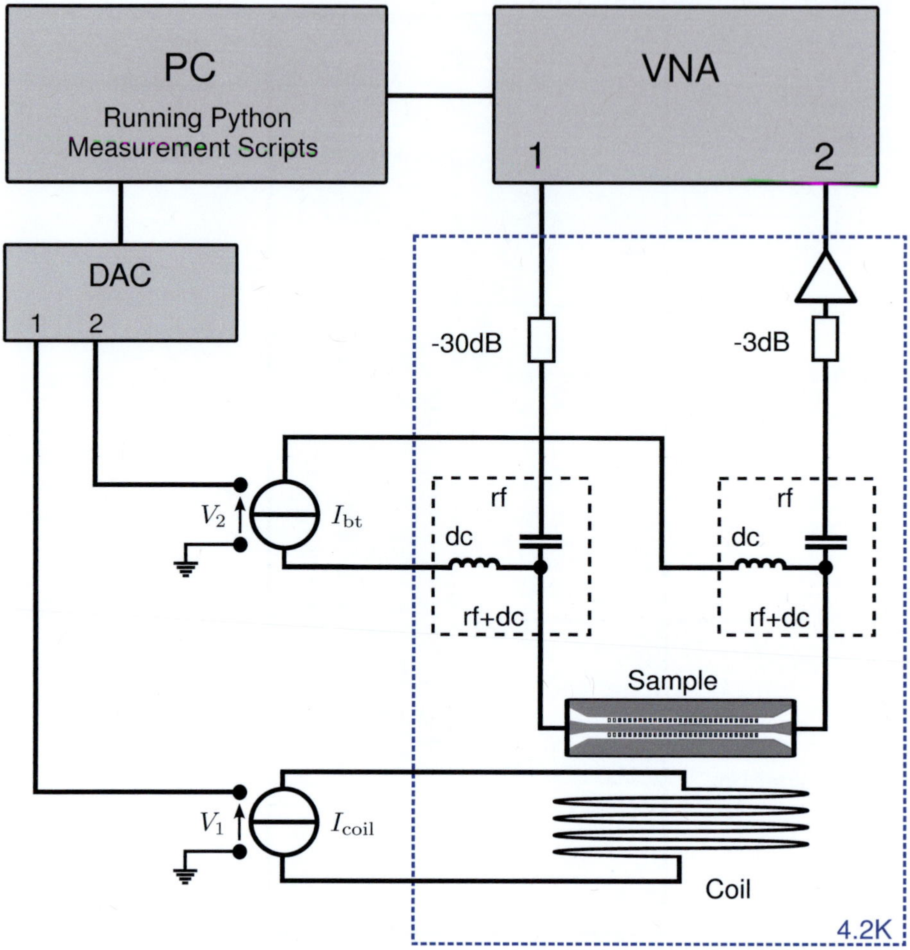

Figure 3.9.: Sketch of the measurement setup. The PC manages the experiment by controlling the VNA and a *digital to analog converter (DAC)*. Two current sources, that are controlled by the DAC, supply the currents for the coil I_{coil} and the bias-Ts I_{bt}. The blue dashed line indicates the cryogenic part of the setup. It consists of an input attenuator, the two bias-Ts (indicated by the black dashed line), the sample, the coil, an output attenuator and a HEMT amplifier as also shown in Fig. 3.8. Note that the orientation of the coil is not correctly shown here. In reality, the generated field is perpendicular to the surface of the sample.

Many of the components introduced in this subsection were developed and fabricated in-house. The sample holder PCB is a modification of an older design and was made in the electronic workshop using a LPKF[5] ProtoMat S100 PCB milling machine. The sample holder was developed as part of a bachelor thesis [Wol12] and is also an improved version of an older design. It was manufactured at the local mechanical workshop, as was the dip-stick on which it is mounted.

A more detailed account of the used devices and instruments and their specifications is given in appendix B.2.

Typical Measurements

Typical measurements for this type of setup are multi-dimensional sweeps controlled by a Python script running on the measurement PC. The innermost sweep parameter (i.e. the one that is swept the fastest) is usually either the frequency or the power of the drive. Sweeping one of these two parameters is left to the network analyzer while the PC collects the resulting one-dimensional traces from the VNA and changes the other parameters as required. This makes it possible to use the calibration routines of the VNA, which is done by replacing the sample-holder box by a through-piece[6] and using the VNA's internal *through-calibration*. The VNA stores the information of the transmitted signal without the sample (i.e. the calibration curve) and uses it to normalize all subsequently measured data. This simplifies the analysis of the data, as now, only the sample holder box and PCB have to be accounted for when modeling the expected transmission through the SQUID-loaded waveguides. The calculations presented in section 3.4 make use of this assumption.

[5] http://www.lpkf.de

[6] In essence, the two cables that would normally connect to the sample holder box are instead connected directly to each other.

3.3. Pulsed setup

For an experiment that will be introduced further on in this thesis (see Sec. 4.4), it is required to apply pulses to the sample in addition to the microwave drive. As the goal of this measurement is to compare the transmission signal before and after applying the pulses, the VNA is set to *continuous wave (CW)* mode. In this setting, it emits a microwave of a single frequency and records S-parameters as a function of time rather than drive frequency or power.

Fig. 3.10 shows the modified setup for this experiment. Instead of connecting the VNA directly to the input of the cryostat, it is connected to the *local oscillator (LO)* port of a microwave mixer at room temperature. The *intermediate frequency (IF)* port of the device is connected to a *pulse/pattern generator (PPG)*. The RF port of the mixer goes to the input port of the cryostat.

Microwave mixers are commonly used to multiply two microwave signals for tasks such as frequency up- or down-conversion or phase detection. When used the way depicted in Fig. 3.10, they can also be used to modulate the amplitude of the signal applied to the LO port. This way, short bursts of microwave power can be created. Later I will refer to these bursts as *rectangular envelope microwave (REM)* pulses.

In this arrangement, the power emitted by the VNA, which drives the diode inside the mixer, has to be at a constant level of about $10\,\mathrm{dBm}$. Due to the limited isolation of the mixer, the minumum power at the rf port of the mixer is about $-25\,\mathrm{dBm}$. Whenever lower power levels were required, more attenuators were installed in addition to the one shown in Fig. 3.10.

The pattern generator is programmed to output a sequence of “low” and “high” voltage levels (shown schematically in Fig. 3.11b), which are applied to the mixer as is the output of the VNA (cf. Fig. 3.11a). The resulting waveform (Fig. 3.11c) is the product of the two inputs. Hence, the low voltage from the PPG leads to a low microwave power, while a high voltage leads to a high power being applied to the sample. In a typical measurement, the computer sends a trigger signal to the PPG, which in turn triggers the VNA to start a measurement and starts to output a low voltage. After a certain time (usually about $0.1\,\mathrm{s}$), the PPG sets the voltage level to high for a brief duration ($2-200\,\mathrm{ns}$).

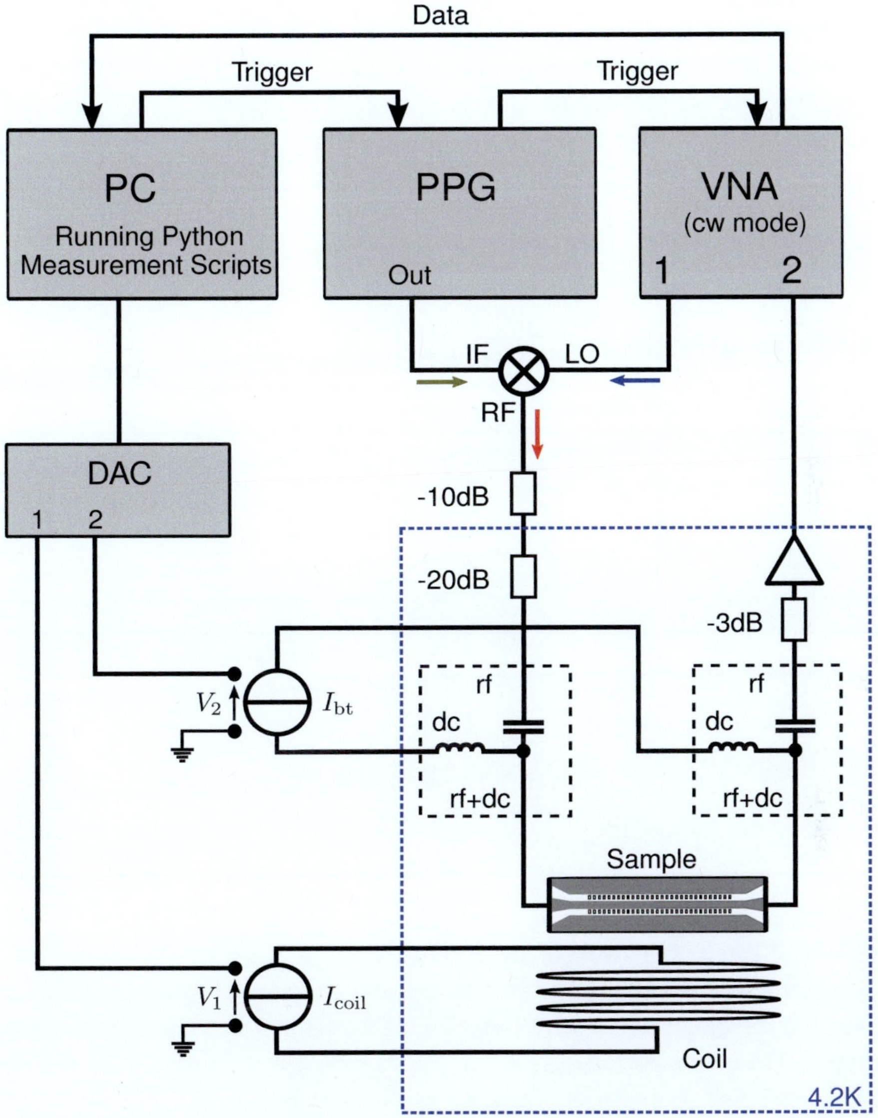

Figure 3.10.: Sketch of the measurement setup for pulsed measurements. The low temperature part is almost identical to the original design. On the input side of the room temperature part, the VNA is set to continuous wave mode and its output is mixed with that of a *pulse/pattern generator (PPG)* before being fed to the sample. The colors of the arrows correspond to the colors in Fig. 3.11, which shows the signals coming from the different devices.

Finally, the voltage is set back to the initial low level for the rest of the measurement (usually about $0.9\,\mathrm{s}$). The VNA records the transmitted microwave signal as a function of time throughout this sequence. The measurement is usually repeated a number of times to get some statistical data.

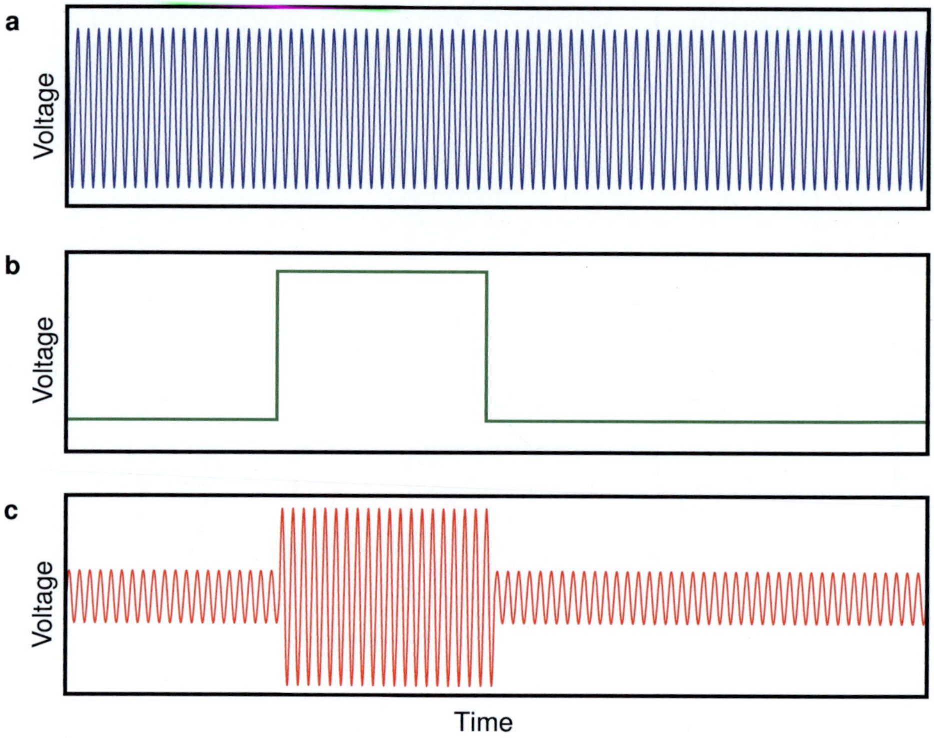

Figure 3.11.: Sketch of the waveforms coming from: **a**, The VNA. **b**, The pulse/-pattern generator. **c**, The output of the mixer is a constant microwave signal with a rectangular envelope microwave pulse. This last signal is what is applied to the sample. The colors correspond to the color of the arrows in Fig. 3.10, which show the respective propagation direction of the signals.

3.4. Modeling

In order to be able to compare the different results obtained using the methods outlined in this chapter, a connection between theoretically accessible quantities and measurable S-parameters has to be formed. In the following, I will therefore introduce a method that can be used to calculate the scattering parameters of the innermost part of the setup as a function of the SQUID's flux susceptibility. This fundamental quantity can be obtained from the steady state solution of the SQUID's equation of motion (cf. Eq. (2.34)).

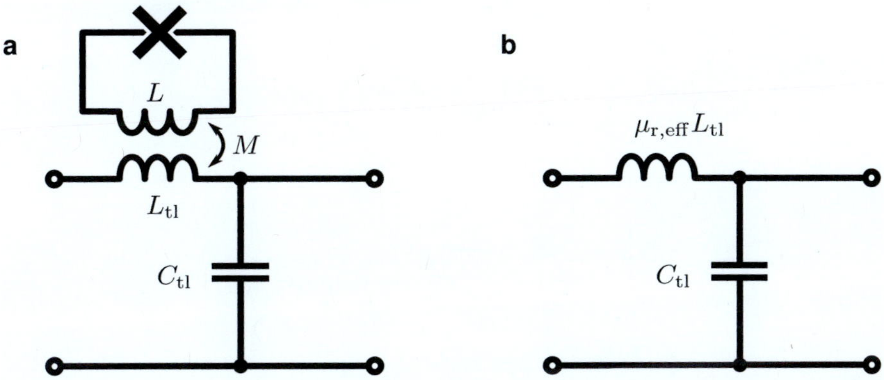

Figure 3.12.: Equivalent circuit of a CPW unit cell (represented as a transmission line) containing a SQUID. **a**, The SQUID is coupled to the CPW by a mutual inductance M. **b**, The same scenario can be represented as a pure TL section where the value of the inductance is modified by the presence of the SQUID (cf. Eq. (3.5)).

In case of a single SQUID embedded in one of the gaps of a coplanar waveguide, we first have to find an electric circuit equivalent of the CPW-SQUID unit cell. Fig. 3.12a shows such a unit cell. The CPW is represented as a lossless transmission line section of length l with characteristic inductance L' and capacitance C' per unit length. The inductance of the SQUID loop L is coupled to the CPW inductance $L_{\mathrm{tl}} = L'l$ via a mutual inductance

$$M = k\sqrt{L\,L_{\mathrm{tl}}}, \tag{3.3}$$

where k is called the coupling coefficient. The impedance of the TL inductance is modified by the presence of the SQUID because in addition to the voltage due to the current in the transmission line I_{tl}, another voltage is induced by the screening currents flowing in the SQUID loop I_{sq}. Making the usual harmonic assumption in the complex notation we can write the current in the transmission line and the total flux through the SQUID as $I_{\mathrm{tl}} = I_0 \exp(i\omega t)$ and $\Phi = \Phi_a \exp(i\omega t + \delta)$, respectively. Taking into account that the current in the TL causes the external driving flux of the SQUID via their mutual inductance $\Phi_{\mathrm{ext}} = M \cdot I_{\mathrm{tl}}$ and that screening current, total flux and external flux are connected by Eq. (2.4), we can write the voltage drop across the inductance as

$$\begin{aligned} V_{\mathrm{lsq}} &= L_{\mathrm{tl}}\frac{dI_{\mathrm{tl}}}{dt} + M\frac{dI_{\mathrm{sq}}}{dt} = L_{\mathrm{tl}}\frac{dI_{\mathrm{tl}}}{dt} + \frac{M}{L}\frac{d(\Phi - \Phi_{\mathrm{ext}})}{dt} \\ &= i\omega L_{\mathrm{tl}}\left(1 + k^2\left(\frac{\Phi_a}{\Phi_{\mathrm{ea}}}e^{i\delta} - 1\right)\right) I_0 e^{i\omega t}. \end{aligned} \tag{3.4}$$

The modified impedance of the inductance is then

$$Z_{\mathrm{lsq}} = \frac{V_{\mathrm{lsq}}}{I_{\mathrm{tl}}} = \underbrace{i\omega L_{\mathrm{tl}}}_{Z_{\mathrm{L,tl}}}\underbrace{(1 + k^2\chi_\Phi)}_{\mu_{\mathrm{r,eff}}}, \tag{3.5}$$

where $Z_{\mathrm{L,tl}}$ is the impedance of the transmission line inductance without the SQUID and χ_Φ the flux susceptibility that was introduced earlier. This complex quantity contains the amplitude ratio and phase difference between the flux oscillations in the SQUID and the drive. From this last equation it can be seen that the influence of the SQUID on the inductance can be considered as a change of the effective, relative permeability $\mu_{\mathrm{r,eff}}$ seen by the TL inductance. This is also indicated in Fig. 3.12b.

According to Ref. [Poz05], the ABCD matrix (introduced in Sec. 2.2.4) of the unit cell is then

$$\tilde{A}_{\mathrm{stl}} = \begin{pmatrix} 1 - \mu_{\mathrm{r,eff}}\omega^2 L_{\mathrm{tl}}C_{\mathrm{tl}} & \mu_{\mathrm{r,eff}} i\omega L_{\mathrm{tl}} \\ i\omega C_{\mathrm{tl}} & 1 \end{pmatrix} \tag{3.6}$$

At this point we can calculate the Z matrix (Eq. (2.20)) and from that the S matrix (Eq. (2.16)) of the unit cell. Due to the nature of the setup, however, it is impossible to calibrate the measurements just before and after the SQUID-loaded unit cell. The closest we can move the calibration planes to the sample are the SMP connectors on the PCB (cf. Fig. 3.5). For this reason, we have to include everything in between those two connectors into the model if we want to be able to compare it to actual measurements. Starting from one connector, this means a section of CPW on the PCB of length l_1 with matched impedance $Z_0 = 50\,\Omega$ and propagation constant β_1, then the bond connection between PCB and chip, then another section of CPW on the chip (length $l_2 \approx 2\,\mathrm{mm}$, $Z_0 = 50\,\Omega$ and propagation constant β_2) on each side of the unit cell. Using the ABCD matrices $\tilde{A}_1$, $\tilde{A}_{\mathrm{bond}}$ and $\tilde{A}_2$ for the first CPW section, the bond wires and the second CPW section, respectively, the complete ABCD matrix for the network between the connectors is given by

$$\tilde{A}_{\mathrm{c}} = \tilde{A}_1 \cdot \tilde{A}_{\mathrm{bond}} \cdot \tilde{A}_2 \cdot \tilde{A}_{\mathrm{stl}} \cdot \tilde{A}_2 \cdot \tilde{A}_{\mathrm{bond}} \cdot \tilde{A}_1. \tag{3.7}$$

Here, the CPW matrices are represented by pieces of lossless transmission line

$$\tilde{A}_1 = \begin{pmatrix} \cos(\beta_1 l_1) & iZ_0 \sin(\beta_1 l_1) \\ i\sin(\beta_1 l_1)/Z_0 & \cos(\beta_1 l_1) \end{pmatrix}, \tag{3.8a}$$

$$\tilde{A}_2 = \begin{pmatrix} \cos(\beta_2 l_2) & iZ_0 \sin(\beta_2 l_2) \\ i\sin(\beta_2 l_2)/Z_0 & \cos(\beta_2 l_2) \end{pmatrix}, \tag{3.8b}$$

while the bond wires are modeled as an inductive in-line scatterer with inductance L_{bond}

$$\tilde{A}_{\mathrm{bond}} = \begin{pmatrix} 1 & i\omega L_{\mathrm{bond}} \\ 0 & 1 \end{pmatrix}. \tag{3.9}$$

Finally, using Eq. (2.20) and Eq. (2.16) we can calculate the complete scattering matrix from $\tilde{A}_{\mathrm{c}}$ (Eq. (3.7)) which can be compared to measurement result that have been calibrated at the SMP connectors on the PCB.

In reality, of course, all of the components that have been assumed as lossless are lossy as well. This will lead to a difference between measurements and calculation that, however, is sufficiently small to not obscure the qualitative results.

For the sake of completeness, it should be mentioned that the presented method can also be extended to the case of a CPW loaded with many SQUIDs (such as the one in Fig. 3.3) although this will not be needed in the following. For N unit cells, one has to cascade the ABCD matrices $\tilde{A}_{\mathrm{N}} = \tilde{A}_{\mathrm{stl}}^N$ and reduce the distance l_2 by half the length of the newly introduced array. This scenario has been studied in detail in Ref. [But14]. If there is more than one SQUID per unit cell (such as in G2-2 or G2-54 where both CPW gaps are filled with SQUIDs), Eq. (3.4) has to be modified by adding the voltage drop caused by the additional SQUID(s).

3.5. Rectangular Waveguide Setup

The two-dimensional SQUID arrays of the third generation were measured in a setup, which is completely different from the one introduced so far. As these samples do not contain their own CPWs, they have to be inserted into an external waveguide. This is achieved by placing them in the center of a rectangular K_u band waveguide as shown in Fig. 3.13, which is similar to the method described in Ref. [Tre+13].

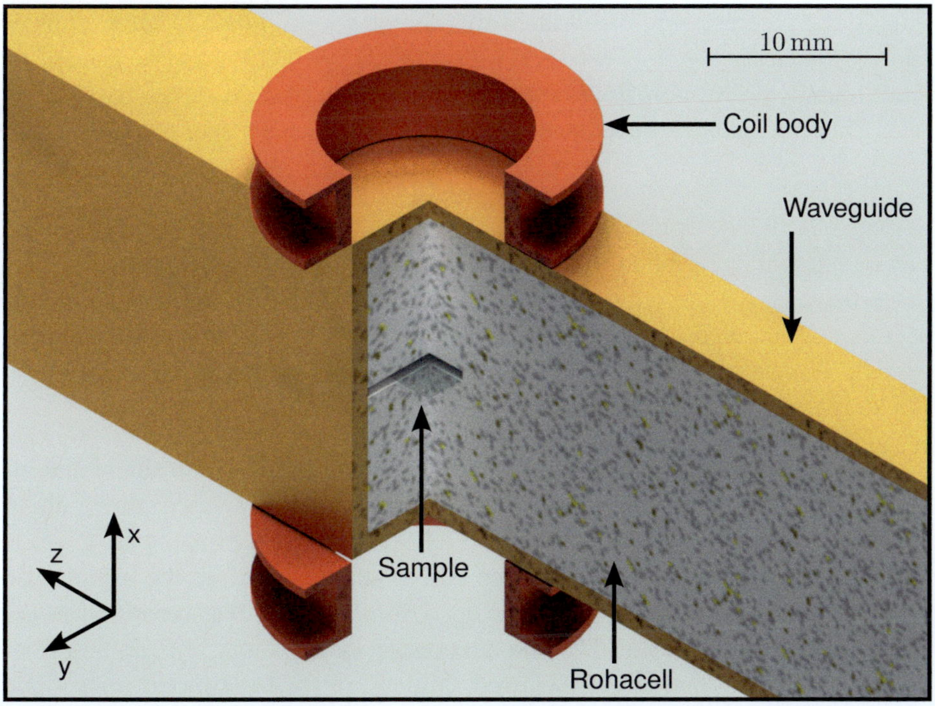

Figure 3.13.: Computer drawing of the rectangular waveguide setup (three quarter section view). The rectangular waveguide is filled with Rohacell which holds the chip (semi-transparent block sticking out of the Rohacell) in place at the center of the structure. A split coil (only body shown) is mounted on the waveguide.

To achieve the best possible coupling to the magnetic field, the chip containing the SQUIDs has to be positioned with the SQUID-loops perpendicular to the x-axis[7] of the waveguide. To hold it in place while influencing the waveguide properties as little as possible, the chip is inserted into a piece of Rohacell[8] which completely fills the interior of a section of the waveguide. The Rohacell block is pushed into the waveguide and waveguide to coaxial couplers are installed on either end.

Mounting the sample using this method has both advantages and disadvantages: On the up side, the Rohacell material is almost transparent to the microwave and causes very little reflections. On the downside, thermal contact between sample and waveguide is rather poor. After the waveguide has reached base temperature during a cool-down, it still takes the sample several hours to show signs of superconductivity.
Additionally, positioning the sample inside the waveguide is somewhat inaccurate. This can lead to a difference in effective microwave power or dc bias field across the array.

A magnetic field is applied to the SQUID array through a split coil positioned just outside the waveguide. Due to the geometry of this arrangement (i.e. distance between the coils much larger than their radius) and the positioning inaccuracy mentioned above, the field at the sample has a noticeable gradient over the dimensions of the array which can be seen in some measurements (cf. Fig. 4.14).

The waveguide with couplers is enclosed in a cylindrical μ-metal shield and installed at the base plate of a closed-cycle ^{4}He cryostat at a temperature of about 4.5 K. To improve magnetic shielding even further, the complete base plate is enclosed in a Nb shield, that becomes superconducting before the samples do during cool-down. The rest of the setup is similar to the one described in the previous subsection. There is some attenuation on the input line, an amplifier on the output line and a VNA is used to measure the transmitted microwave signal.

Typical measurements are similar to the previous case, only that here they are controlled by a MATLAB script. Signal levels in this setup are much lower than in a CPW because the coupling between waveguide and SQUID is considerably

[7] Axis are chosen as in Fig. 2.2 and Fig. 3.13.

[8] Rohacell is a rigid, industrial foam with $\mathfrak{Re}(\varepsilon_r)$ close to unity at microwave frequencies and a low loss tangent.

weaker. In the CPW geometry introduced earlier, a single SQUID causes a resonance dip of the order of 1 dB. Here, in the rectangular waveguide experiment, the dip in transmission for an array of 441 SQUIDs is only about 0.1 dB deep.

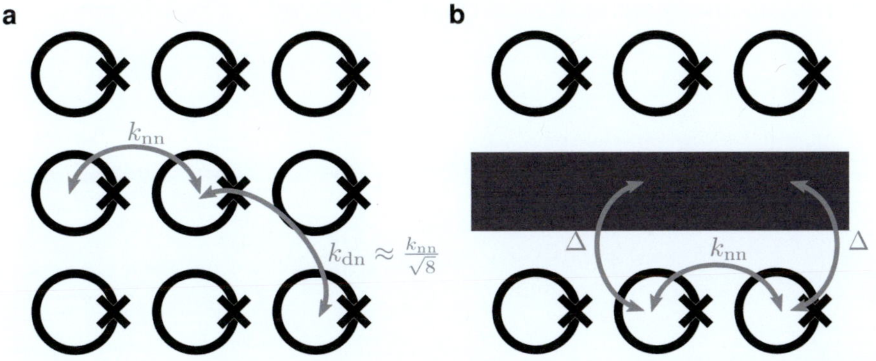

Figure 3.14.: Different coupling scenarios. **a**, In case of a two-dimensional SQUID array in a rectangular waveguide with nearest neighbor coupling of strength $k_{\rm nn}$. **b**, A one-dimensional array in the presence of an additional waveguide (indicated here as the center conductor of a CPW) with nearest neighbor coupling (strength $k_{\rm nn}$) and coupling to the waveguide (strength Δ).

While the weak coupling between waveguide and SQUIDs makes the transmitted signal more challenging to measure, it also makes this setup better suited then the CPW version to investigate the effects of coupling strength in the array. The reason for this can be found when comparing the coupling mechanisms in both scenarios. In the two-dimensional, square lattice case (Fig. 3.14a), the magnetic dipole coupling strength $k_{\rm nn}$ between neighboring SQUID loops has a negative sign. Since $|k_{\rm nn}|$ is usually very small ($|k_{\rm nn}| < 3 \cdot 10^{-2}$), second order effects (of strength $k_{\rm nn}^2$) are of minor concern. When comparing the values of $k_{\rm nn}$ and $k_{\rm dn}$ that were obtained numerically (cf. Sec. 3.1.2), including second order effects (i.e. $k_{\rm dn} = k_{\rm nn}/\sqrt{8} + 2k_{\rm nn}^2 + \mathcal{O}(k_{\rm nn}^3)$) yields slightly more accurate results. If the SQUIDs are embedded in the CPW (Fig. 3.14b), the situation can be completely different. Here, the second order terms (of strength Δ^2) that arise when one SQUID couples to the CPW which in turn couples back to another SQUID, can be larger than the direct (dipole) coupling between neighboring SQUIDs . Since $k_{\rm nn}$ is negative and Δ^2 is positive, the sign of the effective coupling between neighboring SQUIDs depends on the exact geometry (i.e. the distance between neighboring SQUIDs and their position inside the waveguide).

From these considerations, it is clear that the rectangular waveguide is a more suitable setup to investigate the effects of coupling strength in SQUID arrays.

4. Results

"For sufficiently strong drive, the oscillator bifurcates and can occupy one of two metastable states (O_{L} and O_{H}). These states differ significantly in oscillation amplitude and phase."

Vijay et al.

The Josephson bifurcation amplifier [VDS09]

The introductory quote is taken from a review article on Josephson bifurcation amplifiers and highlights the main novelty of this work. In the quoted paper, the authors describe the dynamics of a driven Josephson oscillator by retaining only the first nonlinear term from the series expansion of the Josephson nonlinearity. This essentially reduces the problem to the (analytically solvable) Duffing oscillator discussed in Sec. 2.4.3. While this is a valid approach for their case, we will see that a more general treatment leads to qualitatively different results.

For a sufficiently strong drive, the SQUID does not only become bistable but can develop a large number of stable states. Moreover, we will see how these states can be accessed experimentally and how this makes the SQUID a switchable meta-atom.

4.1. Theoretical analysis

4.1.1. Nonlinearity Beyond the Duffing-Oscillator

Although the description of the SQUID as a Duffing oscillator already provides a number of improvements over the small signal model, there are still some qualitative differences that it does not explain satisfyingly, as we shall see in the following. I will therefore present an approach based on the rotating wave approximation (RWA) that allows for a more accurate description. This approach has been developed in collaboration with M. V. Fistul and M. Marthaler and was published in part in Ref. [Jun+14].

Instead of Taylor-expanding the nonlinearity $\sin\varphi$, we leave this term as it is for now but require the system to oscillate only with the driving frequency ω by assuming

$$\varphi = \varphi_{\mathrm{a}} \cos(\omega t + \delta) = x\cos(\omega t) - y\sin(\omega t), \tag{4.1}$$

with in-phase component $x = \varphi_{\mathrm{a}}\cos(\delta)$, quadrature $y = \varphi_{\mathrm{a}}\sin(\delta)$, amplitude $\varphi_{\mathrm{a}} = \sqrt{x^2 + y^2}$ and phase $\delta = \arctan(y/x)$. This assumption is usually only valid if the frequency of the drive is close to the resonance frequency. To simplify the calculations we will use an equivalent complex notation[1] for the phase φ and the external drive φ_{ext}

$$\varphi = \frac{\varphi_{\mathrm{a}}}{2}\left(e^{i(\omega t+\delta)} + e^{-i(\omega t+\delta)}\right) \equiv b e^{i\omega t} + c.c., \tag{4.2}$$

$$\varphi_{\mathrm{ext}} = \frac{\varphi_{\mathrm{ea}}}{2}\left(e^{i\omega t} + e^{-i\omega t}\right) = \frac{\varphi_{\mathrm{ea}}}{2} e^{i\omega t} + c.c., \tag{4.3}$$

with

$$b = x/2 + iy/2. \tag{4.4}$$

In the next steps we will make some assumptions for the other terms in the rf-SQUID equation of motion (Eq. (2.34)) requiring each of them to only retain spectral components that oscillate with ω. We first calculate the derivatives of φ while allowing for a slow evolution of b on the timescale $\tau = 1/\lambda \gg 2\pi/\omega$.

$$\dot{\varphi} = \dot{b} e^{i\omega t} + i\omega b e^{i\omega t} + c.c. \tag{4.5}$$

$$\ddot{\varphi} = \ddot{b} e^{i\omega t} + 2i\omega\dot{b} e^{i\omega t} - b\omega^2 e^{i\omega t} + c.c. \tag{4.6}$$

[1] We will sometimes only calculate the terms corresponding to $+\omega$ and denote their complex conjugate counterparts (belonging to $-\omega$) with $c.c.$

Later, we will require the time dependence of b to vanish for a steady state. Making use of the slow variation of b ($\dot{b} \propto \lambda$, $\ddot{b} \propto \lambda^2$), we know that the terms containing the time derivatives are small (because λ is small) so that we only have to keep the ones proportional to $\omega\dot{b}$:

$$\dot{\varphi} \simeq i\omega b e^{i\omega t} + c.c. \tag{4.7}$$
$$\ddot{\varphi} \simeq 2i\omega\dot{b}e^{i\omega t} - b\omega^2 e^{i\omega t} + c.c. \tag{4.8}$$

Now, we find a suitable representation of the $\sin\varphi$ term. To this end, we expand it into a Fourier series

$$\sin(\varphi) = \sum_{n=-\infty}^{\infty} c_n e^{in\omega t}, \tag{4.9}$$

with

$$c_n = \frac{1}{T}\int_{t_0}^{t_0+T} \sin(\varphi)\, e^{-in\omega t}\, dt. \tag{4.10}$$

In this expression, we only keep the components oscillating with $\pm\omega$ and the stationary term namely $n = 0, \pm 1$. The full calculation of $c_{\pm 1}$ can be found in appendix A.1. Here, we will use the result which is

$$c_0 = 0, \tag{4.11}$$
$$c_1 = J_1(2|b|)\frac{b}{|b|}, \tag{4.12}$$
$$c_{-1} = c_1^*. \tag{4.13}$$

Hence,

$$\sin(\varphi) \simeq c_1 e^{i\omega t} + c_{-1}e^{-i\omega t} = \frac{b\, J_1(2|b|)}{|b|}e^{i\omega t} + c.c. \tag{4.14}$$

Here and later, J_n is the n-th order Bessel function of the first kind. We can now insert the results from equations (4.2), (4.7), (4.8) and (4.14) into Eq. (2.34). The resulting equation can be separated into two parts[2] of which, for simplicity, we will only consider the one proportional to $\exp(i\omega t)$.

$$b + \frac{i\omega\beta_\mathrm{L}}{\omega_\mathrm{c}}b + \frac{\beta_\mathrm{L}}{\omega_\mathrm{p}^2}\left(i\omega 2\dot{b} - b\omega^2\right) + \frac{\beta_\mathrm{L} b}{|b|}J_1(2|b|) = \frac{1}{2}\varphi_\mathrm{ea}. \tag{4.15}$$

[2] One being proportional to $\exp(i\omega t)$, the other one to $\exp(-i\omega t)$.

We now solve for $\dot{b}$ and require it to vanish for a stationary solution.

$$\frac{2i\omega\beta_{\mathrm{L}}}{\omega_{\mathrm{p}}^2}\dot{b} = \frac{1}{2}\varphi_{\mathrm{ea}} - b - i\beta_{\mathrm{L}}\frac{\omega}{\omega_{\mathrm{c}}}b + \beta_{\mathrm{L}}\frac{\omega^2}{\omega_{\mathrm{p}}^2}b - \frac{\beta_{\mathrm{L}}b}{|b|}J_1(2|b|) \stackrel{!}{=} 0. \tag{4.16}$$

Using Eq. (4.4), this can be written as two real equations that we will henceforth refer to as the *equations of motion in the rotating frame*:

$$0 = x\left(1 - \beta_{\mathrm{L}}\frac{\omega^2}{\omega_{\mathrm{p}}^2}\right) - \beta_{\mathrm{L}}\frac{\omega}{\omega_{\mathrm{c}}}y + 2\beta_{\mathrm{L}}J_1(\varphi_{\mathrm{a}})\frac{x}{\varphi_{\mathrm{a}}} - \varphi_{\mathrm{ea}}, \tag{4.17a}$$

$$0 = y\left(1 - \beta_{\mathrm{L}}\frac{\omega^2}{\omega_{\mathrm{p}}^2}\right) + \beta_{\mathrm{L}}\frac{\omega}{\omega_{\mathrm{c}}}x + 2\beta_{\mathrm{L}}J_1(\varphi_{\mathrm{a}})\frac{y}{\varphi_{\mathrm{a}}}, \tag{4.17b}$$

or, in a more compact form,

$$0 = \frac{\partial g}{\partial x} - \beta_{\mathrm{L}}\frac{\omega}{\omega_{\mathrm{c}}}y, \tag{4.18a}$$

$$0 = \frac{\partial g}{\partial y} + \beta_{\mathrm{L}}\frac{\omega}{\omega_{\mathrm{c}}}x, \tag{4.18b}$$

where g is the potential or *quasi-energy* of the system:

$$g = \frac{1}{2}\left(x^2 + y^2\right)\left(1 - \beta_{\mathrm{L}}\frac{\omega^2}{\omega_{\mathrm{p}}^2}\right) - \varphi_{\mathrm{ea}}x - 2\beta_{\mathrm{L}}J_0\left(\sqrt{x^2 + y^2}\right) \tag{4.19}$$

Stability of the solutions

Equations (4.18a) and (4.18b) still yield two qualitatively different types of solutions: The ones that are stable under an infinitesimal perturbation, and the ones that are not and can therefore not be observed experimentally. This is related and similar to the unstable solutions found for the Duffing oscillator (cf. dashed line in Fig. 2.9). To distinguish between those two cases, we compute the Jacobian (or stability matrix) of the system:

$$M_{\mathrm{J}} = \begin{pmatrix} \frac{\partial \dot{x}}{\partial x} & \frac{\partial \dot{x}}{\partial y} \\ \frac{\partial \dot{y}}{\partial x} & \frac{\partial \dot{y}}{\partial y} \end{pmatrix}. \tag{4.20}$$

The eigenvalues of this matrix evaluated at the fixed points given by the solutions of equations (4.18a) and (4.18b) carry information about the stability of the system [RHL77]. These so called local Lyapunov exponents quantify whether two points in phase space[3] around the investigated solution would converge upon each other or not. If this is the case, the real part of all eigenvalues is negative and the solution is stable. In all other cases, the solution is unstable.

A general, one-dimensional solution

When the previously stated equations of motion are solved numerically, one can find a number of simultaneously stable solutions for moderately strong driving around the resonance frequency. Surprisingly, all solutions found with this procedure, lie on circles in the x-y-plane. These circles cross the origin and are centered at a point on the y-axis. This indicates that the solutions may be described by a single parameter (the angle of the circle) instead of two. Excluding the cases $x = 0$ and $y = 0$, we can rewrite the problem in new coordinates ϵ and ϑ

$$x = \epsilon \sin \vartheta, \tag{4.21a}$$
$$y = \epsilon \left(\cos \vartheta - 1 \right), \tag{4.21b}$$

with $\epsilon \in \mathbb{R}$ and $\vartheta \in (0, 2\pi)$. This leads to

$$\varphi_{\mathrm{a}} \left(\vartheta \right) = |\epsilon| \sqrt{2 - 2 \cos \vartheta}, \tag{4.22a}$$
$$\delta \left(\vartheta \right) = -\frac{\vartheta}{2}. \tag{4.22b}$$

We then divide Eq. (4.17a) by x and Eq. (4.17b) by y and subtract them from each other leading to

$$\frac{A}{x} - \frac{B}{y} = 0, \tag{4.23}$$
$$\beta_{\mathrm{L}} \frac{\omega}{\omega_{\mathrm{c}}} \left(\frac{x^2 + y^2}{y} \right) + \varphi_{\mathrm{ea}} = 0, \tag{4.24}$$

where we defined A and B as the right hand sides of Eq. (4.17a) and Eq. (4.17b), respectively. If we now insert the new coordinates we get

$$\epsilon = \frac{\varphi_{\mathrm{ea}}}{2\beta_{\mathrm{L}}} \frac{\omega_{\mathrm{c}}}{\omega}. \tag{4.25}$$

[3] In this case, phase space refers to the rotating frame (i.e. $x, \dot{x}, y, \dot{y}$.)

This defines the radius of the circle on which the solutions lie. Next, we evaluate the sum

$$\frac{A}{x} + \frac{B}{y} = 0, \tag{4.26}$$

leading to

$$2\left(1 - \beta_\mathrm{L}\frac{\omega^2}{\omega_\mathrm{p}^2}\right) + \beta_\mathrm{L}\frac{\omega}{\omega_\mathrm{c}}\left(\frac{x^2 - y^2}{xy}\right) + 4\beta_\mathrm{L}\frac{J_1(\varphi_\mathrm{a})}{\varphi_\mathrm{a}} - \frac{\varphi_\mathrm{ea}}{x} = 0. \tag{4.27}$$

By multiplying this expression with $\varphi_\mathrm{a}^2/(4\beta_\mathrm{L}\epsilon^2)$ and using the new coordinates, we can write it as

$$\underbrace{\left(\frac{1}{\beta_\mathrm{L}} - \frac{\omega^2}{\omega_\mathrm{p}^2}\right)(1 - \cos\vartheta) - \frac{\omega}{\omega_\mathrm{c}}\sin\vartheta + \frac{\sqrt{2 - 2\cos\vartheta}}{\epsilon}J_1(\epsilon\sqrt{2 - 2\cos\vartheta}) = 0.}_{=C_\mathrm{ss}(\vartheta)} \tag{4.28}$$

This is the *one-dimensional steady state condition in the rotating frame.*

To come to this conclusion we previously excluded the cases $x = 0$ and $y = 0$. We will now check the generality of this assumption starting with the case $y = 0$. From Eq. (4.17b), we can immediately see that this implies $x = 0$. Eq. (4.17a), however, can only be fulfilled with $x = 0 \wedge y = 0$ if $\varphi_\mathrm{ea} = 0$, which is not the case meaning that no soulution with $y = 0$ exists unless there is no driving. Next, $x = 0$ inserted into Eq. (4.17a) imposes the requirement $y = -2\epsilon$. This case is valid under the new coordinates because Eq. (4.17b) and Eq. (4.28) are identical for $x = 0$, $y = -2\epsilon$ and $\vartheta = \pi$. We can therefore state that the new coordinates do not bring about a loss of generality and the newly found one-dimensional solution is in fact a general solution for the problem.

As for the two-dimensional case, the Eq. (4.28) also yields stable and unstable solutions which can be distinguished using the method outlined above. From comparison with numerics, it was found that the stable ones seem to correspond to the ones with a positive slope $\partial C_\mathrm{ss}/\partial\vartheta > 0$ while the other solutions with $\partial C_\mathrm{ss}/\partial\vartheta \leq 0$ are unstable.

4.1.2. Analysis of the Theoretical Model

In this subsection, I will analyze the implications of the formalism that was developed above by comparing it to the known limits of weak and strong driving as well as to the Duffing model. I will also discuss the predictions of our model for those regions of the parameter space that cannot be described by the other models.

Limits of Weak and Strong Driving

To verify that this solution is valid in both the limit of weak and strong driving, we can approximate the Bessel functions in the corresponding regime. For weak driving, the oscillations in the SQUID and thus the argument of J_1 will be small, so we can approximate it by the first term of its series expansion

$$J_1(x) = \frac{x}{2} + o(x^3). \tag{4.29}$$

Using this in Eq. (4.28) yields the expressions

$$\varphi_\mathrm{a} = \left(\left(\beta_\mathrm{L}\frac{\omega}{\omega_\mathrm{c}}\right)^2 + \left(1-\beta_\mathrm{L}\frac{\omega^2-\omega_\mathrm{p}^2}{\omega_\mathrm{p}^2}\right)^2\right)^{-\frac{1}{2}}, \tag{4.30}$$

$$\delta = -\arctan\left(\beta_\mathrm{L}\frac{\omega}{\omega_\mathrm{c}}\left(1-\beta_\mathrm{L}\frac{\omega^2-\omega_\mathrm{p}^2}{\omega_\mathrm{p}^2}\right)^{-1}\right). \tag{4.31}$$

They are equivalent to the ones expected for the linearized weak driving case [LL98] (see section 2.4.2).

In the case of very strong driving, the last term on the left hand side of Eq. (4.28) becomes negligibly small as it scales roughly with $1/\varphi_\mathrm{ea}$.

The resulting expressions for amplitude and phase are then

$$\varphi_\mathrm{a} = \left(\left(\beta_\mathrm{L}\frac{\omega}{\omega_\mathrm{c}}\right)^2 + \left(1-\beta_\mathrm{L}\frac{\omega^2}{\omega_\mathrm{p}^2}\right)^2\right)^{-\frac{1}{2}}, \tag{4.32}$$

$$\delta = -\arctan\left(\beta_\mathrm{L}\frac{\omega}{\omega_\mathrm{c}}\left(1-\beta_\mathrm{L}\frac{\omega^2}{\omega_\mathrm{p}^2}\right)^{-1}\right). \tag{4.33}$$

As expected, these equations are identical to those of the RLC circuit (see section 2.4.2). In conclusion, we can say that the model correctly reproduces the linear weak- and strong-driving limits.

Number of Stable Solutions

It was already hinted at above that Eq. (4.28) can have more than one stable solution depending on the SQUID parameters and the driving. Fig. 4.1 shows the number of stable solutions in dependence on frequency and amplitude of the drive.

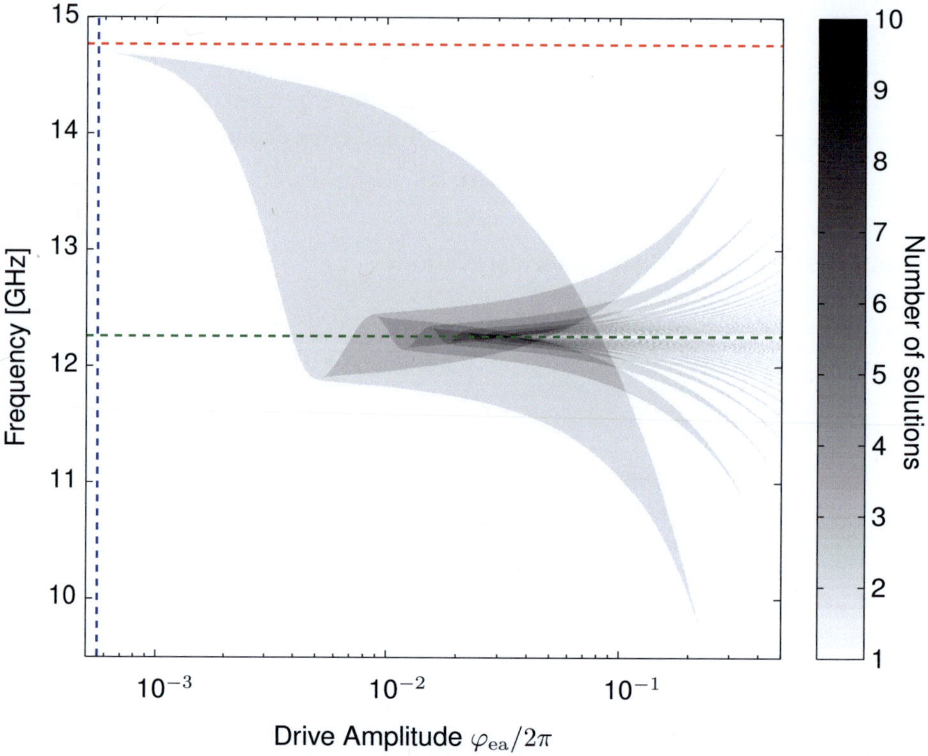

Figure 4.1.: Number of stable solutions of Eq. (4.28) for different values of the frequency and amplitude of the drive. The number is color coded with white being one and darker colors corresponding to larger numbers of stable solutions. The red, green and blue lines correspond to the weak-driving resonance frequency f_r, the LC resonance frequency f_{r0} and the critical driving amplitude φ_{crit}, respectively.

Coming from the weak driving side (left), where the system has only one stable state for all frequencies, the first sign of bistability can be seen close to

$f_r \approx 14.77\,\mathrm{GHz}$ (red dashed line). The bistable region then broadens and moves to lower frequencies as the amplitude of the drive increases. A third state appears close to the LC resonance frequency $f_{r0} \approx 12.26\,\mathrm{GHz}$ (green dashed line) followed by more and more states adding up to a maximum of 10 simultaneously stable solutions for this choice of parameters. When increasing φ_{ea} even further, this number eventually goes down again, oscillating between one and two for a while until finally going back to one for very large driving amplitudes (not shown here).

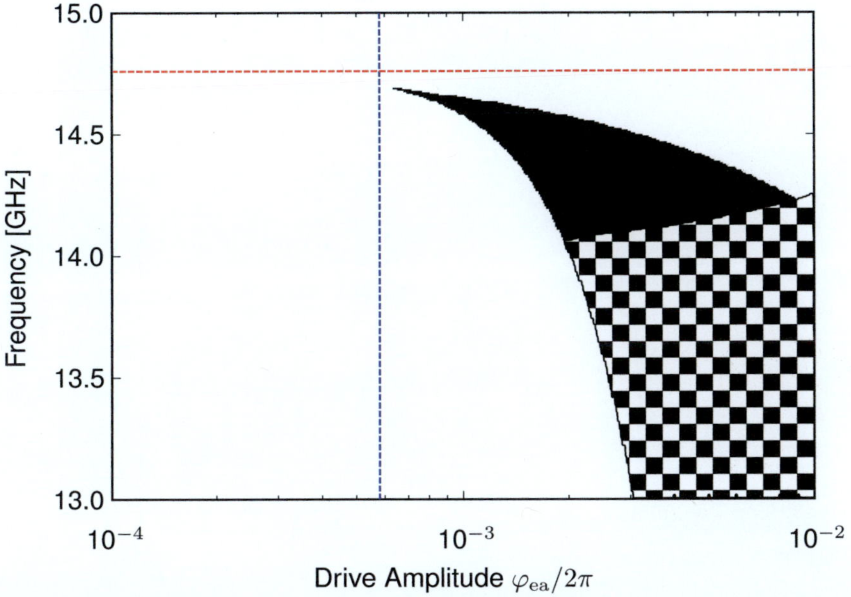

Figure 4.2.: Number of stable states of the SQUID using the Duffing-oscillator model as a function of the driving amplitude and frequency. In the white region Eq. (2.48) has only one stable solution, in the black region the system is bistable and the checkered area indicates where the underlying approximation (cf. Eq. (2.47)) does not hold any more.

For the onset of bistability, this can be nicely compared to the results of the Duffing-oscillator model that is shown in Fig. 4.2 (blue dashed line). According to the Duffing oscillator model (cf. Eq. (2.49)), the critical driving amplitude above which the oscillator becomes bistable is

$$\frac{\varphi_{\mathrm{crit}}}{2\pi} \approx 5.6 \cdot 10^{-4}. \tag{4.34}$$

This number is consistent with the results of both Figs. 4.1 and 4.2. Generally, both figures agree well with each other for the range of parameters in which the φ^3 approximation for the $\sin(\varphi)$ term of Eq. (2.47) holds.

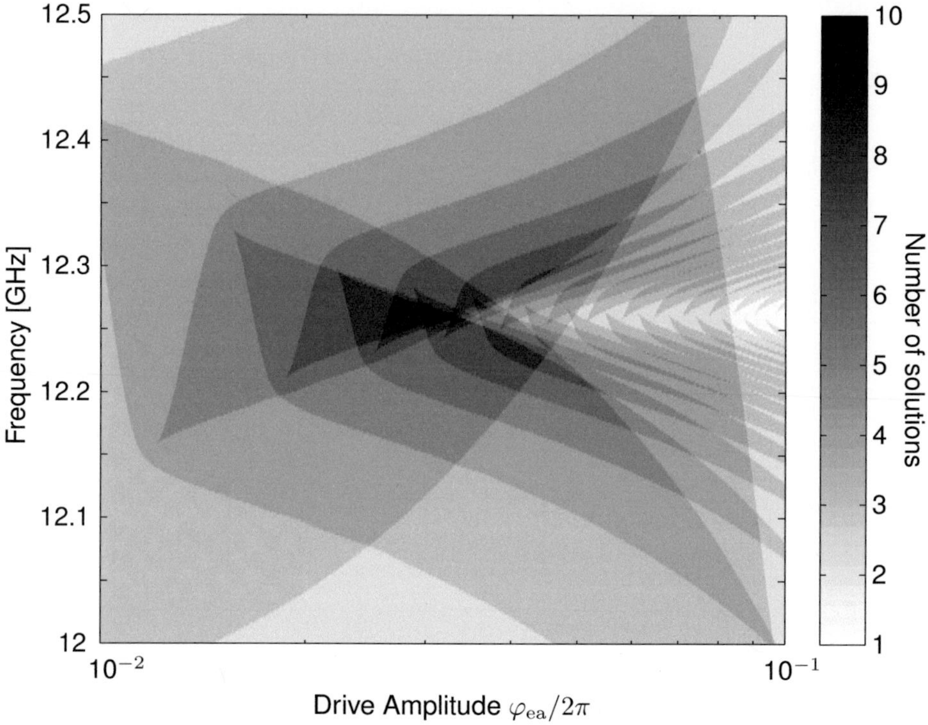

Figure 4.3.: Number of stable solutions (color coded) of Eq. (4.28) for different values of the frequency and amplitude of the drive. Zoom to the "interesting" region (cf. Eqns. 4.35a & 4.35b) of Fig. 4.1.

From Fig. 4.1 and the zoomed in version shown in Fig. 4.3, it can be seen that the most "interesting" region (in the sense that it has the highest number of states) is confined around the LC resonance frequency f_{r0} and a narrow range in driving amplitudes. The limits of this range can be deduced from the last term of C_{ss} (cf. Eq. (4.28)). Its prefactor has to be larger than that of the other terms ($1/\epsilon \gg \omega/\omega_c$) and its argument such that J_1 oscillates faster in ϑ than the other terms ($\epsilon \gg 1$). Thus, we can identify the relevant criteria as

$$\omega \approx \frac{\omega_p}{\sqrt{\beta_L}}, \tag{4.35a}$$

$$\frac{2\sqrt{\beta_L}\omega_p}{\omega_c} \ll \varphi_{ea} \ll 2\beta_L. \tag{4.35b}$$

Influence of the Damping

The presented results are all based on a fixed set of SQUID parameters close to the ones of the actual samples. While the response of the SQUID (and its range of multi-stability) depends on all of the design parameters, the damping plays a particularly interesting role.

While the flux dependence of the resonance line in the weak driving limit is essentially independent of the damping, the range of multi-stability and the number of stable solutions are severely affected when the damping (through R and thus ω_c) is changed. As the resistor R in the RCSJ model of the junction is connected in parallel to the other network elements (cf. Fig. 2.5b), $R \propto \omega_c$ is inversely proportional to the damping. Thus, larger values of R correspond to a lower damping and vice versa.

The dependence of the number of stable solutions of Eq. (4.28) on the damping at the strong driving resonance frequency f_{r0} is illustrated in Fig. 4.4. The most characteristic features of this plot can be understood qualitatively from the equations introduced earlier. The number of solutions increases with R, which can be explained using the same logic that was used in the previous subsection to derive the criteria for the "interesting" region (cf. Eq. (4.35b)). In order for Eq. (4.28) to support more solutions, the argument of the J_1 term has to be such that the term oscillates faster in ϑ than the other terms. Since this argument is proportional to $\epsilon \propto \omega_c \propto R$, smaller damping leads to a larger number of solutions. Simultaneously, the onset of multi-stability at small driving amplitudes shifts to lower φ_{ea} values for increasing R while the upper limit is almost independent of the damping. This, as well, is consistent with the rough approximation of the "interesting" region derived earlier, where the lower limit depends on $\omega_c \propto R$ while the upper limit is constant.

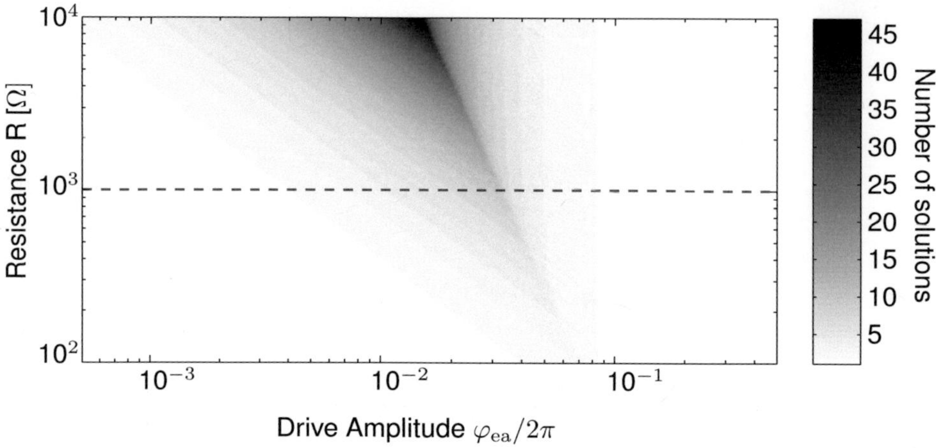

Figure 4.4.: Number of stable solutions (color coded) of Eq. (4.28) for different values of the drive amplitude and the RCSJ resistance at a fixed drive frequency $f = f_{r0}$. The red line indicates the R value used in the other plots in this section. Several trends are visible for increasing R values: First, the maximum number of solutions increases. Second, the onset of multi-stability shifts to smaller drive amplitudes. Third, the range of powers for which the system is multi-stable broadens.

This fact can be used to derive a criterion for a lower bound of R for which Eq. (4.35b) can no longer be satisfied. Evaluated for the actual parameters it requires that $R \gg \Phi_0/(2\pi I_{\mathrm{c}}\sqrt{LC}) \approx 14\Omega$ for the oscillator to be multi-stable at $f = f_{\mathrm{r0}}$. This value is of the same order of magnitude as the actual value found by evaluating Eq. (4.28) ($\approx 30\Omega$). From this we can conclude that, while the exact shape of the multi-stable region strongly depends on the damping, the effect itself is quite robust. It persists even for damping values than are much larger than the ones used in the experiment.

Validity of the Rotating Wave Approximation

Until now, the model has been developed under the assumption that the oscillations in the loop have the same frequency as the drive (cf. Eq. (2.46)). This is not intuitively clear as Josephson junctions and SQUIDs are known to support other phase cycles (e.g. period doubling) and chaos [Kau83]. Unfortunately, there is no simple criterion that would allow us to exclude such a scenario analytically. In rf-SQUIDs, however, these phenomena are usually associated with a multi-well potential ($\beta_\mathrm{L} > 1$) [RS83; FBK83; RS84]. In this work, on the other hand, the potential is not only single-welled ($\beta_\mathrm{L} \approx 0.5$), but resembles an only slightly perturbed parabola (cf. black curve in Fig. 2.7a). Furthermore, no evidence of higher period cycles or chaos has been found in numerical simulations in the parameter range in question.

To further underline the validity of the chosen approximations and thus the applicability of the model, we can calculate to what extent higher harmonics contribute to the phase evolution of the oscillator. For this purpose, we will use the more general form of the phase

$$\varphi = \sum_{n=1}^{\infty} \left[x_n \cos(n\omega t) + y_n \sin(n\omega t)\right] , \tag{4.36}$$

(cf. Eq. (2.45)) now keeping all terms up to $n = 3$. The quasi energy (cf. Eq. (4.19)) then takes the form

$$\begin{aligned} g \;=\; & \frac{1}{2}\left(1 - \beta_\mathrm{L}\frac{\omega^2}{\omega_\mathrm{p}^2}\right)(x_1^2 + y_1^2) - \varphi_\mathrm{ea}x_1 - 2\beta_\mathrm{L}J_0(\sqrt{x_1^2 + y_1^2}) \\ & + \sum_{n=2,3}\left[\frac{1}{2}\left(1 - \beta_\mathrm{L}\frac{n^2\omega^2}{\omega_\mathrm{p}^2}\right)(x_n^2 + y_n^2)\right. \\ & \left. -\beta_\mathrm{L}J_n(\sqrt{x_1^2 + y_1^2})\sin(\pi n/2)\left(x_n\cos(n\delta) - y_n\sin(n\delta)\right)\right] . \end{aligned} \tag{4.37}$$

Here, we assumed that x_n and y_n remain small for $n > 1$. The equations of motion then read

$$\beta_\mathrm{L}\frac{n\omega}{\omega_\mathrm{p}^2}\dot{y}_n = -\frac{\partial g}{\partial x_n} - \beta_\mathrm{L}\frac{n\omega}{\omega_\mathrm{c}}y_n , \tag{4.38}$$

$$\beta_\mathrm{L}\frac{n\omega}{\omega_\mathrm{p}^2}\dot{x}_n = \frac{\partial g}{\partial y_n} - \beta_\mathrm{L}\frac{n\omega}{\omega_\mathrm{c}}x_n . \tag{4.39}$$

From Eq. (4.37), it can be seen that the 2ω oscillations do not couple to the fundamental ones, so only the 3ω mode has to be accounted for. Its amplitude can be calculated as

$$\varphi_a^{(3)} = \sqrt{x_3^2 + y_3^2} = \frac{\beta_\mathrm{L} \left| J_3\left(\varphi_a^{(1)}\right)\right|}{\sqrt{\left(1 - 9\beta_\mathrm{L}\frac{\omega^2}{\omega_\mathrm{p}^2}\right)^2 + 9\beta_\mathrm{L}^2\frac{\omega^2}{\omega_\mathrm{c}^2}}}. \tag{4.40}$$

To verify the initial assumption, we have to assert that

$$\frac{\varphi_a^{(3)}}{\varphi_a^{(1)}} \ll 1. \tag{4.41}$$

The left-hand side of this expression can be estimated by a few simple assumptions: First, we find an estimate for the upper limit of the Bessel function divided by its argument: $J_3(x)/x < 1/9\,\forall\, x$. Next, we assume that $\beta_\mathrm{L} < 1/2$ and $9\beta_\mathrm{L}^2\omega^2/\omega_\mathrm{c} \ll 1$. Finally, on resonance, we have $\beta_\mathrm{L}\omega^2/\omega_\mathrm{p}^2 = 1$, so $\varphi_a^{(3)}/\varphi_a^{(1)} < 1/144$. A better estimate for the frequency range of interest is given by the weak driving tunability range bounded from below and above by

$$\omega_\mathrm{p}\sqrt{\beta_\mathrm{L}^{-1} - 1} < \omega < \omega_\mathrm{p}\sqrt{\beta_\mathrm{L}^{-1} + 1}. \tag{4.42}$$

Even in this case the ratio between 3rd and 1st order oscillations remains small: $\varphi_a^{(3)}/\varphi_a^{(1)} < 1/63$. Fig. 4.5 shows the maximum of $\varphi_a^{(3)}/\varphi_a^{(1)}$ for different values of the driving amplitude and frequency. It shows that our estimations are applicable and underlines the validity of Eq. (4.41).

4.1.3. Comparison with Numerical Simulations

The previous considerations showed that the newly developed model is consistent with what we already know from the simplified models. Ultimately, measurements will have to show whether the expected behavior does indeed occur. For now, however, we will take an intermediate step by comparing our model to results of numerical simulations. By solving the rf-SQUID equation of motion (cf. Eq. (2.34)) as a function of time, we can directly compare the resulting steady state of amplitude and phase to the solutions obtained from Eq. (4.28).

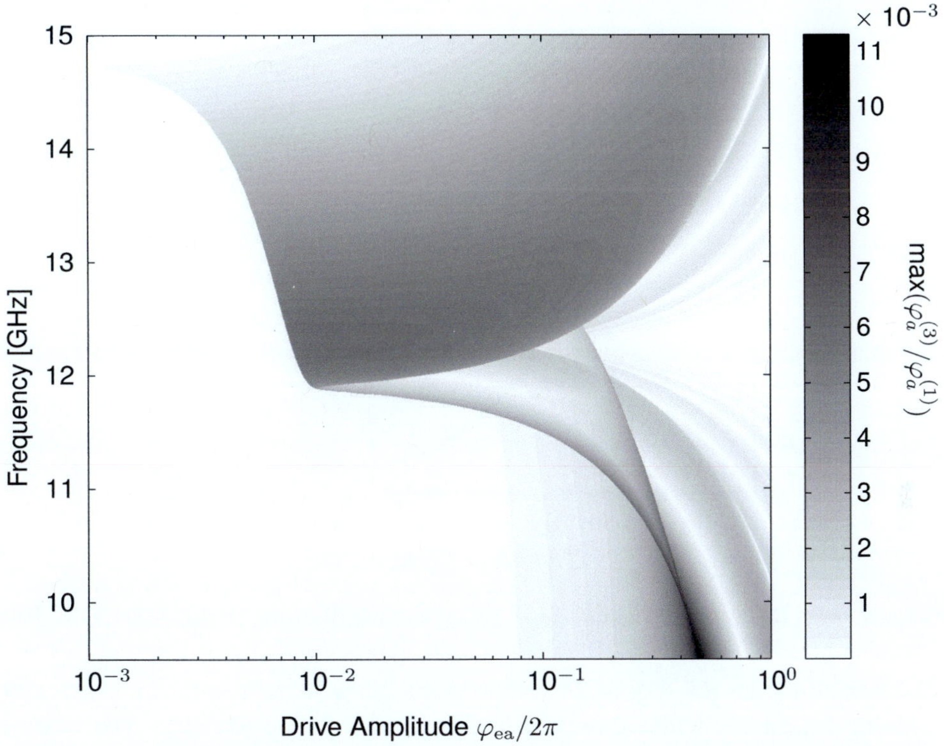

Figure 4.5.: Maximum ratio between 3rd and 1st harmonic oscillations as a function of the drive amplitude and frequency. Maximum means that in the multi-stable regions, the expression $\varphi_a^{(3)}/\varphi_a^{(1)}$ was evaluated according to Eq. (4.40) for all states and the largest value plotted.

There are several ways how simulations can be performed in order to find the different steady states of Eq. (2.34). One example is running the simulation several times for the same drive but with varying initial conditions. Here, we chose a different approach, which will later allow us to compare both analytical and numerical results to measurements.

We start by driving the system at a point on the parameter plane (i.e. the φ_{ea}-ω-plane) at which only one stable state exists. We then move adiabatically along a trajectory on the parameter plane (thereby crossing regions of multistability) and back. Depending on the chosen path, we may observe a hysteresis which can be compared to the analytical model.

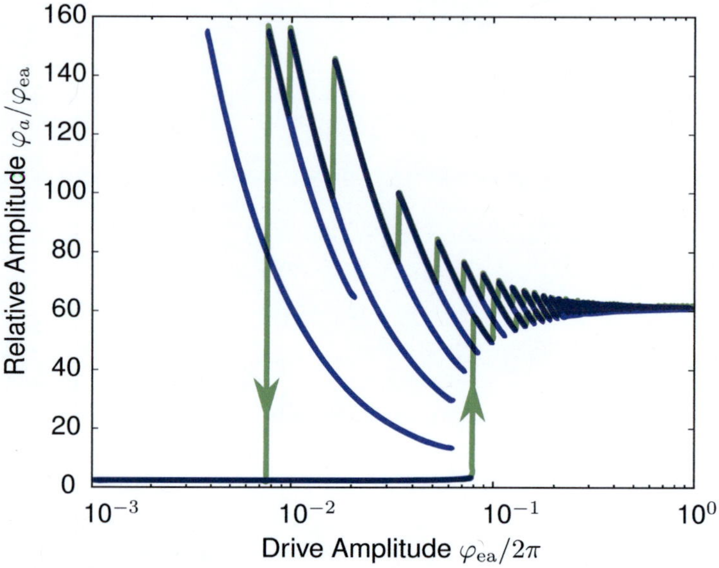

Figure 4.6.: Relative amplitude $\varphi_a/\varphi_{\rm ea}$ of the oscillations in the SQUID at frequency $f = 1.0075 \cdot f_{\rm LC}$. The blue curve shows the stable solutions of the theoretical model (cf. Eq. (4.28)). The green curve is the result of numerically solving Eq. (2.34) while slowly increasing, then decreasing $\varphi_{\rm ea}$. The arrows indicate the direction of the sweep.

In practice, these sweeps are usually done along the $\varphi_{\rm ea}$ axis[4], keeping the frequency at a constant value close to $f_{\rm r0}$. This is due to the fact that the "interesting" region is very confined in frequency but extends over an order of magnitude in the driving amplitude. See Appendix B.1.1 for more details.

Fig. 4.6 shows the amplitudes of both the stable solutions of Eq. (4.28) (blue) and the result of a simulated amplitude sweep (green) for a fixed frequency. The simulation is done by increasing $\varphi_{\rm ea}$ in small steps starting from a value where system has only one stable solution (left in Fig. 4.6). After each step, the system is given enough time for the transient behavior to subside. After reaching the maximum amplitude value, the process is reversed. The resulting data points are shown as the green curve that exhibits a distinct hysteresis.

[4]Unlike the frequency sweeps that are usually used to describe the Duffing oscillator (cf. Fig. 2.9).

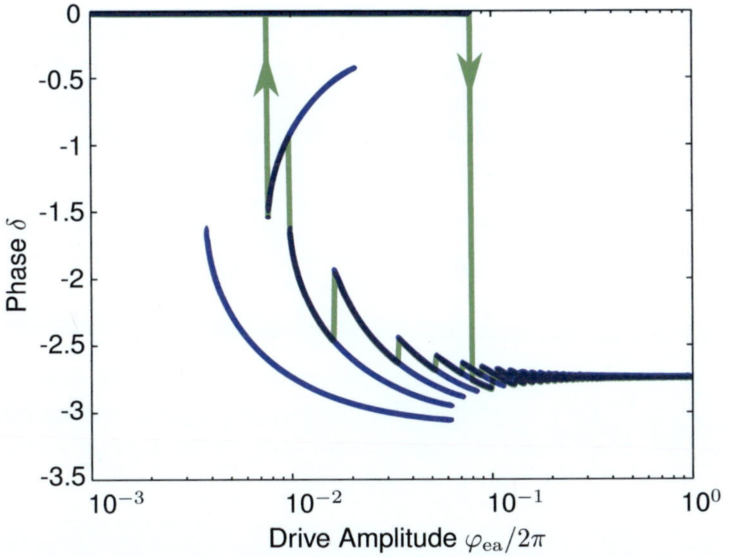

Figure 4.7.: Phase data corresponding to Fig. 4.6.

The numerical solutions follow the stable analytical solutions. Whenever the system is forced to leave one stable state (i.e. when the amplitude is changed such that a previously stable solutions disappears), it jumps to another one.

One unintuitive aspect of Fig. 4.6 is that one solution is not reached by the hysteresis sweep. However, this does not imply that this state does not exist[5]. It is rather an indication that the transient dynamics of the SQUID are more complex than this simple picture might suggest. This becomes even more evident when comparing the hysteresis in amplitude to the corresponding phase (Fig. 4.7). There, it can be seen that during the last jump back to the initial state, the system never actually "passes" the leftmost excited state. This underlines that in such an experiment (sweeping the drive amplitude), it is not always possible to reach all stable states.

To distinguish between the different states, I will from here on use the terms "transparent" and "opaque". In this context, "transparent" refers to states with small φ_{a} like the one on the left because its flux susceptibility is small, meaning

[5]In fact, numerically, the system can be prepared in this state using initial conditions. When the amplitude is swept subsequently, the system remains in the same state.

that it only interacts weakly with an incident wave. "Opaque", on the other hand refers to states that have large values of φ_a. The large flux susceptibility in this case leads to stronger scattering and higher dissipation than in the former case. The latter can easily be seen by calculating the energy W that is dissipated during one oscillation cycle in the junction

$$W = \int_{t_0}^{t_0+T} dt\, P(t) = \frac{1}{R} \int_{t_0}^{t_0+T} dt\, U^2(t). \tag{4.43}$$

Here, $P(t)$ is the power that is dissipated in the RCSJ resistor R at any point in time and $U(t)$ is the voltage over the junction. Using the second Josephson equation (cf. Eq. (1.2b)) and the solution for φ introduced at the beginning of the chapter (cf. Eq. (4.1)), we can calculate W:

$$\begin{aligned} W &= \frac{1}{R}\left(\frac{\Phi_0 \varphi_a \omega}{2\pi}\right)^2 \underbrace{\int_{t_0}^{t_0+T} dt\, \sin^2(\omega t + \delta)}_{=\pi/\omega} \\ &= \frac{\Phi_0^2 \omega}{4\pi R} \varphi_a^2. \end{aligned} \tag{4.44}$$

Consequently, larger oscillation amplitudes φ_a correspond to higher dissipation.

4.2. Measurements in the Weak Driving Regime

In this section, I will present the results of measurements on SQUIDs and arrays thereof in the limit of weak driving. They underline the claim that SQUIDs can be used as low-loss, tunable meta-atoms . Furthermore, measurements in this regime are particularly useful because the quasi-linear weak driving case is well understood and can be described by simple, analytic equations. The data are used to verify design parameters (such as the critical current of the junction or the shunt capacitance) and can serve as a reference for the more complex results shown in sections 4.3 and 4.4.

4.2.1. Single SQUIDs

The first milestone on the road towards Josephson metamaterials was to demonstrate that single rf-SQUIDs can be operated as tunable meta-atoms under the relevant conditions. This included the nature of the coupling (i.e. inductive coupling to a multimode environment like a waveguide) as well as the SQUID design (i.e. small in size, $\beta_\mathrm{L} < 1$ but at frequencies between 10 GHz and 20 GHz).

The first results were obtained on G1-1 (cf. 3.1) and published in Ref. [Jun+13]. Unfortunately, their zero flux resonance frequency lay above 20 GHz and thus outside the available measurement band. One of the measurements is shown in Fig. 4.8. The transmitted power through the waveguide is shown as a function of frequency and constant magnetic flux. The black line is the resonance dip of the SQUID, which shifts in frequency as the magnetic flux is changed. As expected, the curve is is periodic in Φ_0. The central part also features a blue line showing the theoretical curve calculated using Eq. (2.41) and Eq. (2.36). The SQUID parameters I_c, L and C for this curve are given in the table 3.1. As the basic geometry of the SQUID can be very accurately fabricated, its geometric inductance L can be calculated using numerical methods (see appendix B.3 for details). The critical current I_c of the junction was determined from the current to voltage characteristics of a test junction with the same parameters located on the same chip. The only remaining quantity (the capacitance C) cannot be determined through dc measurements and can, in principle, deviate strongly from the design value. This is the case because of the small thickness of the dielectric Nb_2O_5 layer of the shunt capacitor. Its value is therefore obtained by fitting the theoretical dependence of the resonance frequency (cf. Eq. (2.41)) to the data starting from the design value of C.

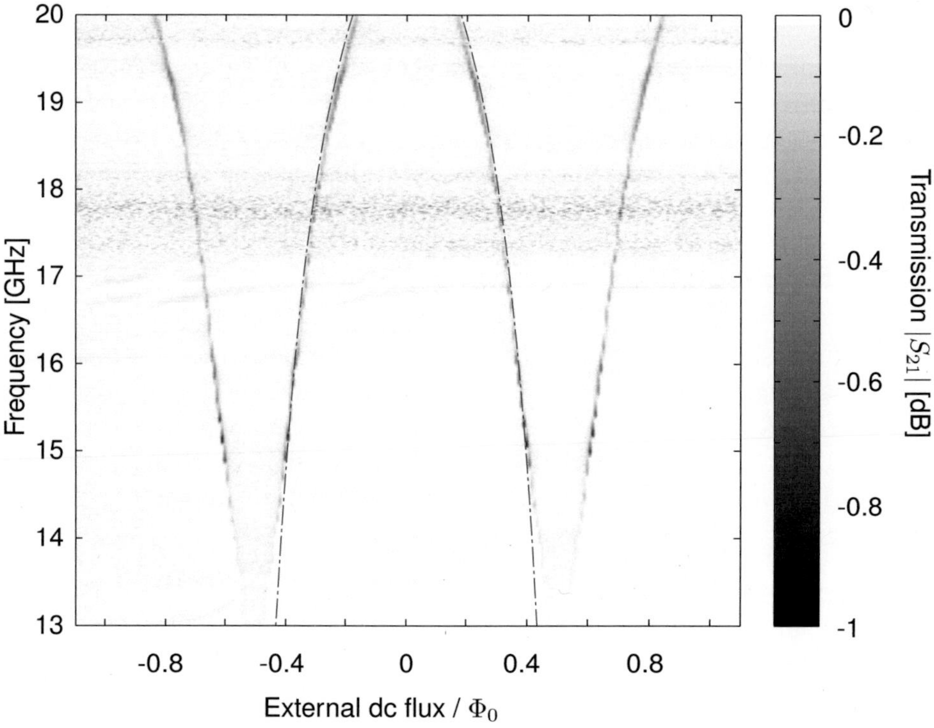

Figure 4.8.: Transmission magnitude of a CPW loaded with a single rf-SQUID meta-atom (G1-1). For clarity, the background has been subtracted. The black line in the data is the resonance dip of the SQUID. Its frequency changes with the externally applied DC flux. The blue dashed line shows the theoretical value of the resonance frequency calculated using Eq. (2.41), Eq. (2.36) as well as the values for this first generation sample.

The remaining discrepancy between theory and experiment is most likely explained by a difference between the critical current of the SQUID and the test junction.

4.2.2. One-Dimensional SQUID Arrays

The next milestone was to move to one-dimensional arrays of SQUID meta-atoms in the coplanar waveguide. This, however, immediately brought to light one of the major challenges that arise when dealing with large numbers of SQUIDs: Their high degree of sensitivity to magnetic field, which is the basis for their tunability, also means that any unwanted, spatially inhomogeneous fields will lead to a distribution of the SQUID resonance frequencies for any given dc flux value. The fields can originate from sources outside the measurement setup if the sample is not perfectly shielded or from magnetic components inside the magnetically shielded setup. The situation becomes even more complex because when cooling the sample through its critical temperature in a magnetic field, Abrikosov vortices can form in the superconducting film. If they are located in one of the SQUID loops, they will contribute more strongly to the flux offset of that particular SQUID.

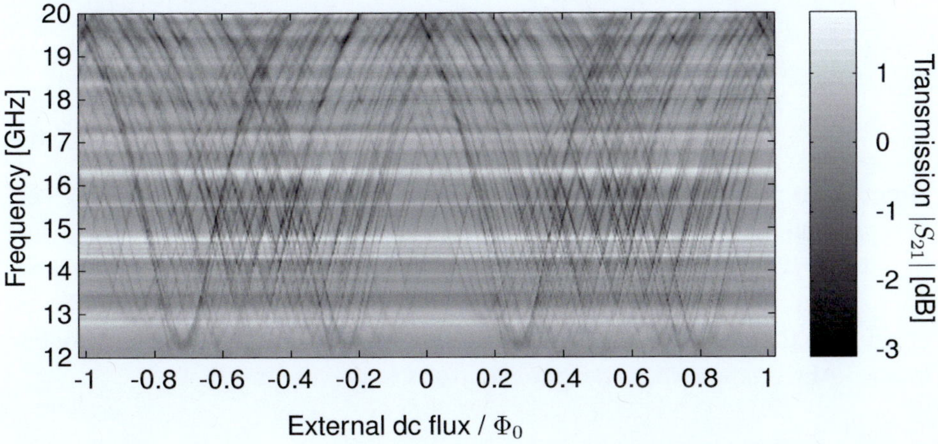

Figure 4.9.: Transmission magnitude through a CPW loaded with 60 first generation SQUIDs (G1-60). The background has been subtracted to enhance visibility of the lines. Positive values of $|S_{21}|$ are a result of this process. A spread in the bias flux values across the array causes the SQUIDs' resonance lines to be offset instead of lying on top of each other.

The first generation SQUIDs were designed in a way which facilitated the formation of vortices. Together with an insufficiently shielded setup also containing several magnetic components, this lead to a very incoherent behavior (cf Fig. 4.9). The distribution in bias fluxes was so large that, instead of one collective resonance curve, each SQUID showed an individual resonance line.

A series of improvements were then made to the sample design and measurement setup in order to achieve a coherent resonance line and bring the sample frequencies down into the measurement band. More details on the optimization process can be found in Refs. [But+13b] and [But14].

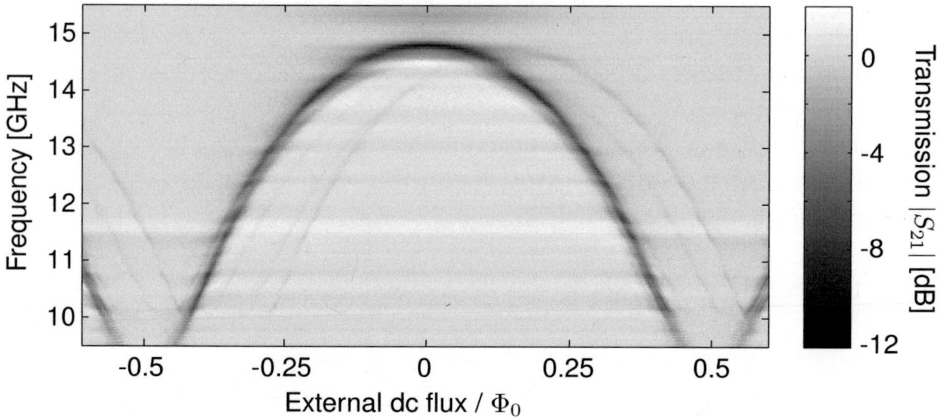

Figure 4.10.: Transmission magnitude through a CPW loaded with 54 second generation SQUIDs (G2-54). The data are normalized to that at a flux of $-\Phi_0/2$. Unlike the data shown in Fig. 4.9, most of the SQUID resonance lines shown here lie on top of each other. A few faint lines can be seen that are offset from the main line. They most likely correspond to SQUIDs that contain trapped Abrikosov vortices despite all efforts to prevent just that.

As a result of the mentioned improvements, the results obtained from the third generation SQUIDs in the optimized setup were much more coherent than before (cf. Fig. 4.10). Instead of many shallow resonance lines, this picture now shows one deep resonance line with only a few outliers.

The data were subsequently used to retrieve an effective, relative permeability of the coplanar waveguide [But+13a]. There, it was shown that $\mu_{\mathrm{r,eff}}$ can deviate strongly from unity, get close to – and for some frequencies also drop below – zero depending on the dc flux value.

4.2.3. Two-Dimensional SQUID Arrays

As the last step before the implementation of a three-dimensional SQUID metamaterial, I will present measurements on two-dimensional arrays. For this purpose, two samples containing square lattices of 21×21 SQUIDs have been included in the third generation of chips.

As such arrays can no longer be incorporated into coplanar waveguides, the measurements were done in a K_u-band rectangular waveguide. This approach has the added advantage that the waveguide can incorporate a stack of such chips thus facilitating the transition from a two-dimensional to a three-dimensional lattice.

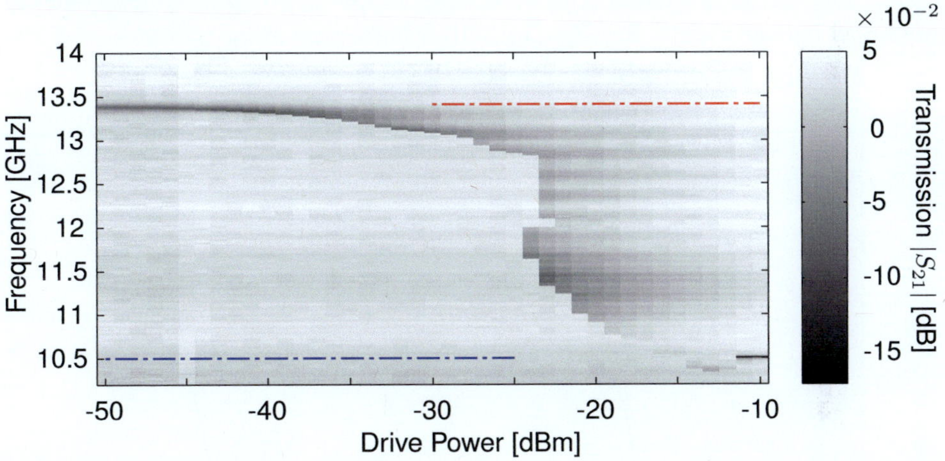

Figure 4.11.: Transmission magnitude ($|S_{21}|$) through a K_u-band rectangular waveguide loaded with 441 strongly coupled SQUIDs (G3-D) as a function of frequency and drive power at zero dc magnetic flux. The data are calibrated to the transmission at $T = 15\,K$ where the SQUIDs are not superconducting. For low driving powers, a clear resonance dip can be seen in the transmission around $f_{\text{max}} = f_{\text{r}}(0) = 13.4\,\text{GHz}$ marked by the red line. No clear resonance is discernible for intermediate drive power levels. At even higher driving power, another resonance dip appears at approximately $f_{\text{r0}} = 10.5$ GHz (blue line). Using these two frequencies and the geometry of the sample, one can calculate the other unknown parameters (I_{c} and C) of the SQUID.

This experiment was also used to investigate the influence of mutual coupling between SQUIDs on the collective behavior. All of the theory and results presented above concerning more than one meta-atom were based on the assumption that there is no direct interaction between them. However, considering the case of a cubic lattice (cf. Eq. (2.7b)), it is clear that to achieve extreme values of μ_r, the spacing d between the meta-atoms has to be minimized, as the magnetic susceptibility is proportional to d^{-3}. Denser spacing, in turn, also increases the mutual interaction between SQUIDs and thus invalidates the assumption of negligible coupling.
Some of the effects that the coupling has on the response to incoming waves can also be found for linear oscillators, like the formation of a magneto-inductive band, that becomes broader with increasing coupling strength. Others, like the interplay between multi-stability and self-organization in the presence of disorder, are unique to this particular system [LT13].

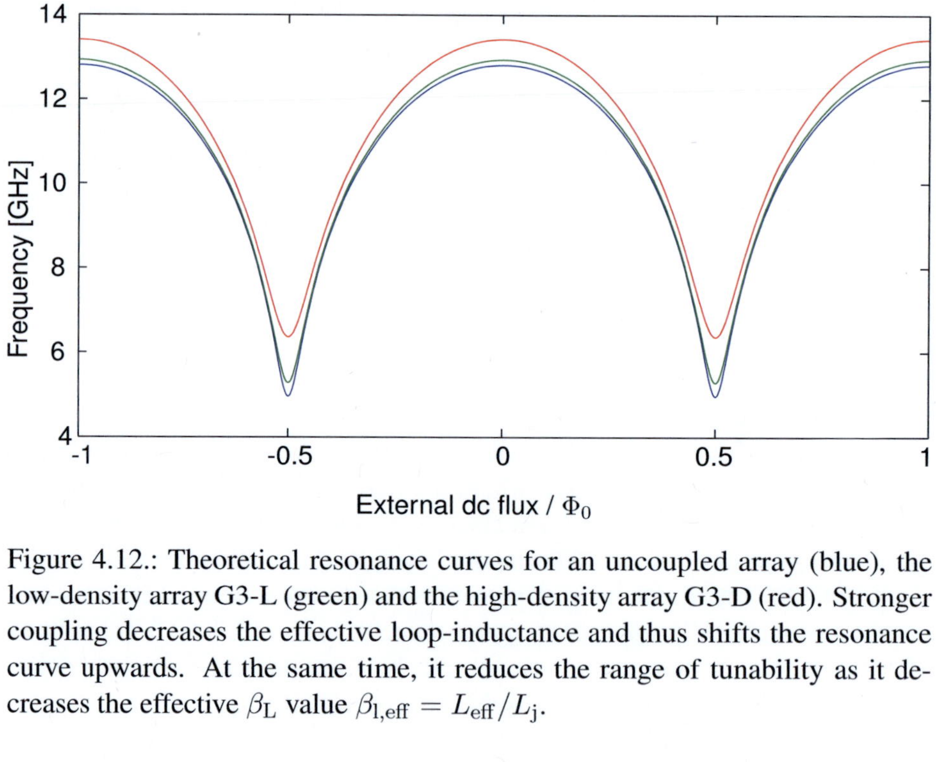

Figure 4.12.: Theoretical resonance curves for an uncoupled array (blue), the low-density array G3-L (green) and the high-density array G3-D (red). Stronger coupling decreases the effective loop-inductance and thus shifts the resonance curve upwards. At the same time, it reduces the range of tunability as it decreases the effective β_L value $\beta_{l,\mathrm{eff}} = L_{\mathrm{eff}}/L_j$.

Here, I will show data from two 2D arrays containing the same number of

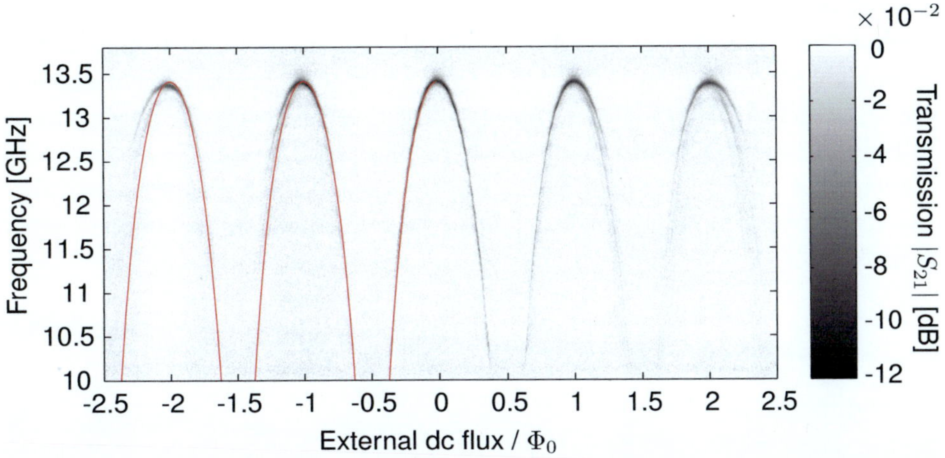

Figure 4.13.: Transmission magnitude ($|S_{21}|$) through a K_u-band rectangular waveguide loaded with 441 strongly coupled SQUIDs (G3-D) as a function of frequency and dc magnetic flux. The background has been subtracted to improve visibility. Note the difference in color scales compared to previous measurements. The resonance curve is visible as a black line. The left half of the image also features the corresponding theoretical resonance curve (cf. Fig. 4.12) as a red line. The lower frequency portion of the resonance curve cannot be measured in this setup as the lower waveguide cutoff frequency is $\approx 9.5\,\text{GHz}$. Note also that the resonance line splits up into two lines at higher flux values.

SQUIDs but with different spacing between the meta-atoms (G3-D and G3-L). The parameters for both arrays can be found in section 3.1 and were used to calculate the theoretical resonance curves. I will use a simplified approach to find the approximate resonance frequency by neglecting the influence of the border of the array and assuming that the coupling can be accounted for by a change of the effective loop inductance through the influence of nearest and second nearest neighbor coupling such that

$$L_{\text{eff}} = L \cdot (1 + 4k_{\text{nn}} + 4k_{\text{dn}})\,. \tag{4.45}$$

Here, we used the coupling strength between one SQUID and its nearest neighbors (k_{nn}) as well as second nearest (diagonal) neighbors (k_{dn}).

The SQUID parameters were recovered from the power dependence of the resonance curve depicted in Fig. 4.11 using the method outlined in Sec. B.1.3. From the known coupling coefficients, one can infer the shape of the resonance curves of an isolated SQUID and the low-density array. Both of them are shown in Fig. 4.12 as the blue and green curve, respectively. The corresponding measurements are shown in Fig. 4.13 and Fig. 4.14 for the high and low-density array, respectively. Both figures also feature the corresponding theoretical curves from Fig. 4.12.

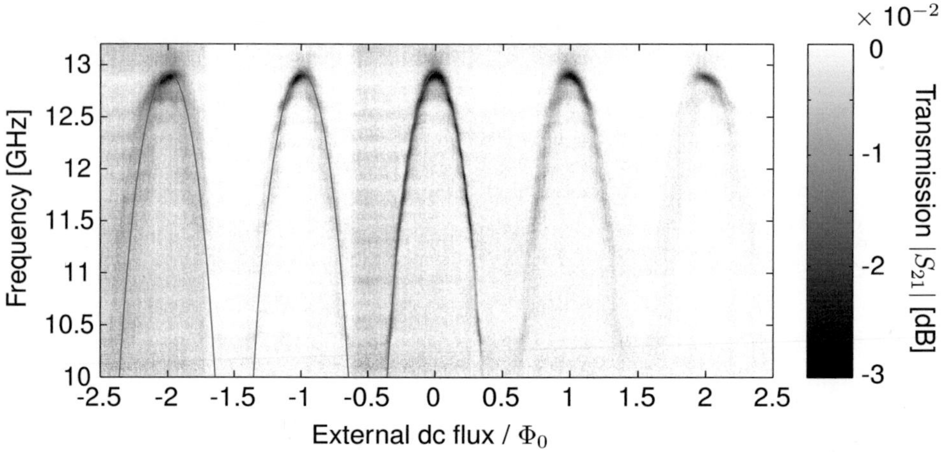

Figure 4.14.: Transmission magnitude ($|S_{21}|$) through a Ku-band rectangular waveguide loaded with 441 weakly coupled SQUIDs as a function of frequency and DC magnetic flux. The background has been subtracted to improve visibility. The resonance feature (black line) is weaker than in the strongly coupled case (cf. Fig. 4.13) but its position is consistent with the theoretical prediction (green line). Note that, unlike in the strongly coupled case, there is no distinct splitting of the resonance feature at higher flux values.

The good agreement between theory and experiment in Fig. 4.13 only underlines that the applied retrieval technique works well. The fact that there is also a very good agreement in Fig. 4.14 is even more interesting as the theoretical curve shown there is only based on the parameters retrieved from the high-density array and the geometry. Knowing that parameters (such as I_c) can vary quite drastically from sample to sample, it cannot be ruled out that the agreement is to some extent coincidental. To strengthen the claim that the theoretical predictions are correct and the agreement is systematic, more measurements would have to be performed. Thus, at this point, we can only claim that the measurements are consistent with the theoretical predictions.

Apart from the frequency dependence and magnitude of the resonance feature, there is another qualitative difference between Fig. 4.13 and Fig. 4.14. For higher values of the external flux, the resonance curve in Fig. 4.14 starts to fan out and becomes fuzzy while the one in Fig. 4.13 splits up into at least two sharp lines.

To understand this effect one has to take the slightly inhomogeneous bias field into account (cf. Sec. 3.5). In case of an uncoupled array, such an inhomogeneity would first lead to a broadening then to a fading of the resonance feature towards higher bias field values similar to what can be seen in the weak coupling measurements (Fig. 4.14). For strongly coupled arrays, the effect can be quite different as subsets of the array may become synchronized [LT13]. Numerical simulations that are not part of this thesis strengthen the suspicion that the split lines in Fig. 4.13 are in fact one indication of synchronization in the array. To verify this, however, one would need to measure the response of the array using a method that gives an insight into the spatial distribution of the SQUIDs' resonance frequencies. Such measurements are currently being prepared using a low temperature laser scanning microscope (LTLSM).

4.3. Multistability and Hysteresis

In this section, I will demonstrate how the existence of multi-stability, which is predicted by theory (see section 4.1.1), can be confirmed experimentally. The most straightforward way of doing so is by slowly varying one parameter of the drive (such as frequency or amplitude) while observing the response of the system.

Similar to the numerical simulations shown in Fig. 4.6, the measurements were done by sweeping the amplitude, i.e. power, of the driving microwave while keeping the frequency constant at values between f_{min} and f_{max}. The reason for this approach rather than the more common frequency sweeps is twofold: First, as already mentioned for the numerical counterpart, the "interesting" region in parameter space extends over several orders of magnitude in driving power but is confined to a narrow frequency band. Second, in experiments, the measurements have to be calibrated. While the frequency dependence of the calibration is very complex and non-monotonic, its power dependence is primarily defined through the nonlinear response of the amplifier which is a slowly-varying, monotonic function of the applied input power.

Hysteresis in a two-SQUID sample

The first measurements of this kind were performed on sample G2-2 (cf. Sec. 3.1) containing two SQUIDs in the coplanar waveguide. The uncalibrated results, shown in Fig. 4.15, exhibit a clear hysteresis curve as predicted by theory. The three curves in this figure correspond to power sweep measurements at different frequencies slightly above the resonance of the strongly driven SQUID $f_{\mathrm{r0}} \approx 12\,\mathrm{GHz}$. The different levels of transmission at low power for those curves illustrate the problem of the frequency dependence of the calibration that was mentioned above. Also, at higher powers, the amplifier starts to become nonlinear which can be seen in the blue curve that starts to bend up at higher powers. Both effects are shown here to emphasize the need for a solid calibration to be able to compare measurements to analytical predictions.

Despite the missing calibration, Fig. 4.15 can already be used to qualitatively verify several of the theoretical predictions. First, the general shape of the hysteresis is mostly consistent with the naive prediction for a scattering meta-atom in a waveguide: Coming from the weak-driving side, the SQUID is in "transparent" state (small flux oscillations) that lead to low scattering and thus *high transmission*. As the power is increased, transmission stays unaltered as long as the SQUID remains in the initial state. As soon as a power is reached at which the state becomes unstable (i.e. -40 dBm for the blue curve), the system settles in a more "opaque" state (larger flux oscillations) that leads to increased scattering and thus *lower transmission*. In the opposite sweep direction, the system follows a different path.

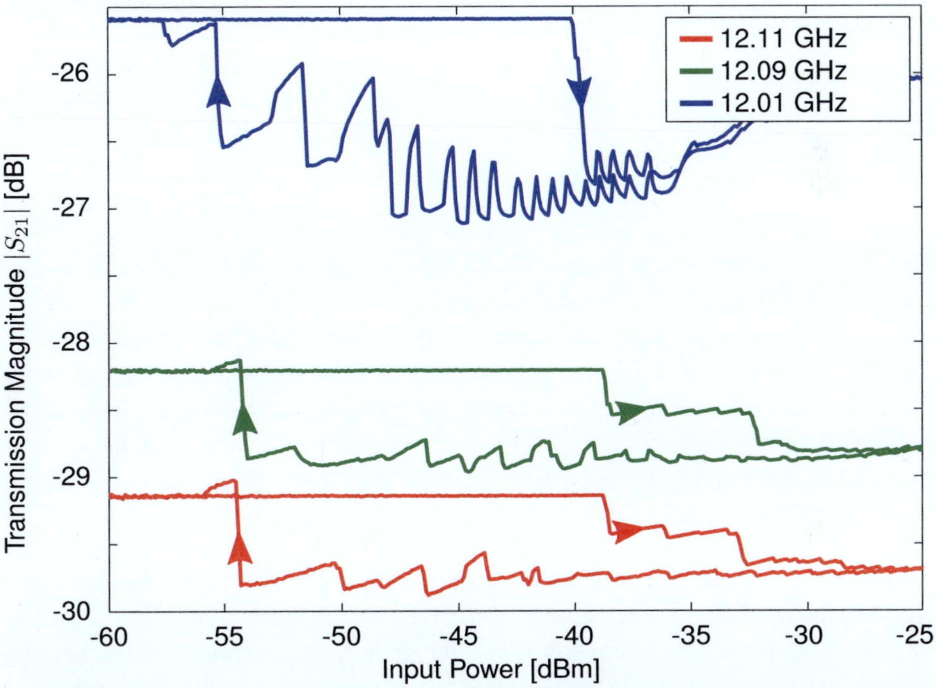

Figure 4.15.: Uncalibrated transmission measurement through a CPW loaded with 2 SQUID meta-atoms (G2-2) while sweeping the drive power at three different frequencies just above the LC resonance frequency $f_{r0} \approx 11.98\,\mathrm{GHz}$. The power was swept in both directions as indicated by the arrows. The different initial levels of the three curves at -30 dBm input power are due to the missing calibration.

Another observation that is consistent with the predictions is that the blue downsweep curve (the one that is closest to f_{r0} in frequency) features a large number of smaller jumps while both the red and green curve only show a few of them. Each of the jumps corresponds to one or both of the SQUIDs jumping from one state to another. This observation is consistent with Fig. 4.3, which predicts a larger number of stable states for drive frequencies close to f_{r0}.

The fact that there are two distinct jumps on the up-sweeps of the green and red curves in Fig. 4.15 indicates that the two SQUIDs jump at different powers. Although this is to be expected for nonzero coupling between the meta-atoms, it complicates the situation and the comparison to theoretical predictions. Consequently, the next measurements were done on sample G2-1 which only contains a single SQUID.

Hysteresis in a single SQUID sample

Fig. 4.16 shows a hysteresis curve (red) for a single SQUID. Additionally, this measurement was thru-calibrated at the microwave connectors on the sample holder PCB to account for the nonlinearity of the amplifier. Up- and down-sweeps (see direction of arrows) both exhibit a number of abrupt jumps between otherwise smooth lines. In most cases, the data coincide with the theoretical prediction that is plotted in gray. The prediction was obtained with the method outlined in appendix B.1.2 and using the same data that was also used in Fig. 4.6 and Fig. 4.7 (blue lines).

At first glance, there is a very good agreement between the measured curve and the theoretical data in a sense that almost all of the measured data lie on top of one of the predicted stable solutions. As with the numerical simulations, the system stays in a state while the power is swept until that state becomes unstable at which point the system settles in another stable state. Unlike the numerical results, the measured data also show noticeable discrepancies. At several points (e.g. at -50 dBm and -40 dBm input power), the measured lines do not follow the predicted path. This could be an effect of the over-simplified coupling model (cf. Sec. 3.4) that was assumed to derive the S-matrix. At other points (e.g. at -53 dBm and -55 dBm input power) the system follows one of the predicted states but jumps to another one at an unexpected power value. While this could also be the effect of incorrect predictions, another possible explanation is that the states are predicted correctly but become unstable in the presence of noise.

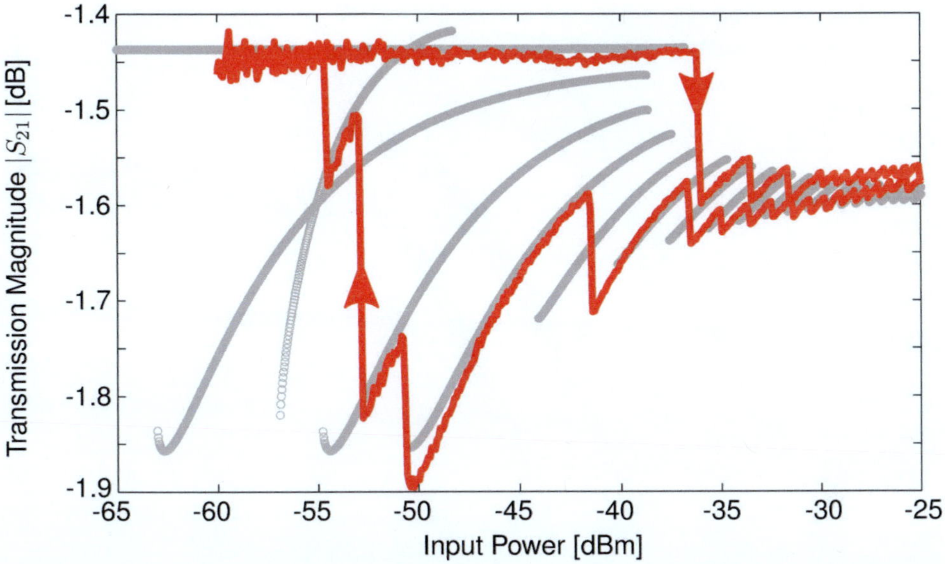

Figure 4.16.: Calibrated transmission measurement (red curve) through a CPW loaded with one SQUID meta-atom (G2-1). The drive power was swept for a fixed frequency of 12.07 GHz ($\approx 1.0075 \cdot f_{\mathrm{r0}}$). Up- and down-sweeps can be distinguished by the direction of the arrows. The gray line shows the theoretical prediction.

One inherent shortcoming of the simple power-sweep experiment presented in Fig. 4.16 is that it only reveals a fraction of the anticipated states. To overcome this, a slightly modified measurement scheme was used. The results can be seen as the blue lines in Fig. 4.17. Each of the blue lines is generated by first initializing the system on a point of the down-sweep (lower red) curve of the hysteresis[6]. From this point, a measurement is started with increasing power, which forces the system to follow whichever state it was occupying, thus revealing more of the anticipated results. To avoid cluttering, all vertical jumps have been removed from the figure. At an input power of -40 dBm, for instance, one can verify the existence of five of the six predicted stable states using this technique.

[6]Performing a partial hysteresis sweep by going to a large power then coming back to a lower power thus following the lower branch of the hysteresis.

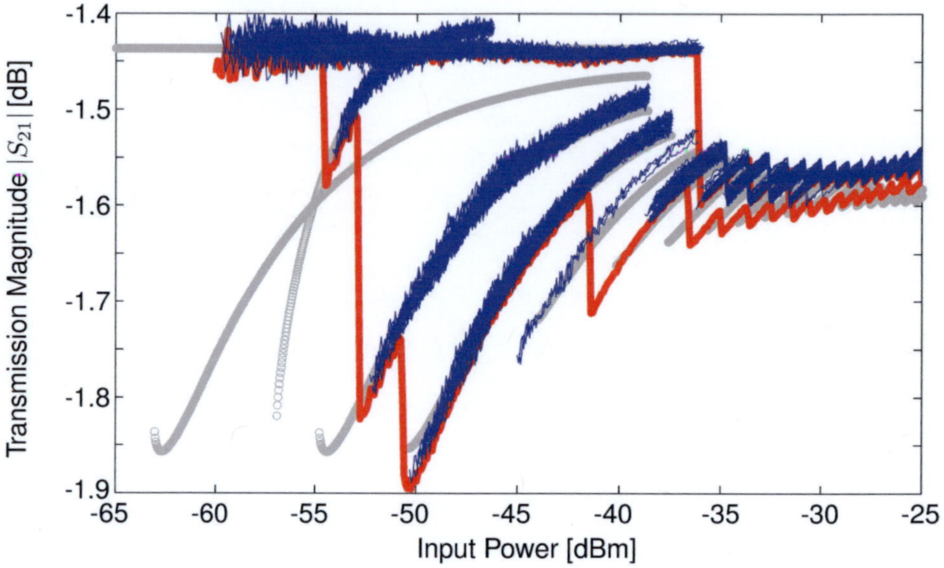

Figure 4.17.: Calibrated transmission measurement (red and blue curves) through a CPW loaded with one SQUID meta-atom (G2-1). The gray and red data are the same as in Fig. 4.16 while the blue data are acquired by successively sweeping the input power starting from various points on the down-sweep (lower red) curve. The characteristic jumps have been removed in the blue data to improve visibility.

The last "missing" state is not seen in either of these experiments. This, however, does not come as a surprise because it was also absent in the simulated power-sweeps (cf. Fig. 4.6 and Fig. 4.7).

Hysteresis in a 54-SQUID sample

As the final part of this section, I will show how multi-stability affects the behavior of samples containing large numbers of SQUID meta-atoms. In particular, I will show measurements on a coplanar waveguide loaded with 54 SQUIDs, two rows of 27, each of them located in one of the CPW's gaps.

From a few simplified considerations it is clear that this situation is much more complex: A single SQUID only has a small number of stable states as a result of Eq. (4.28). In the array, however, the combined state is defined through the combination of the states of all elements. To make matters worse, the effective input power seen by each SQUID in the array also depends on the current state of the rest of the array and vice versa, leading to a self-consistent situation. Despite these effects that complicate modeling, we can still observe a distinct hysteresis in the experiment.

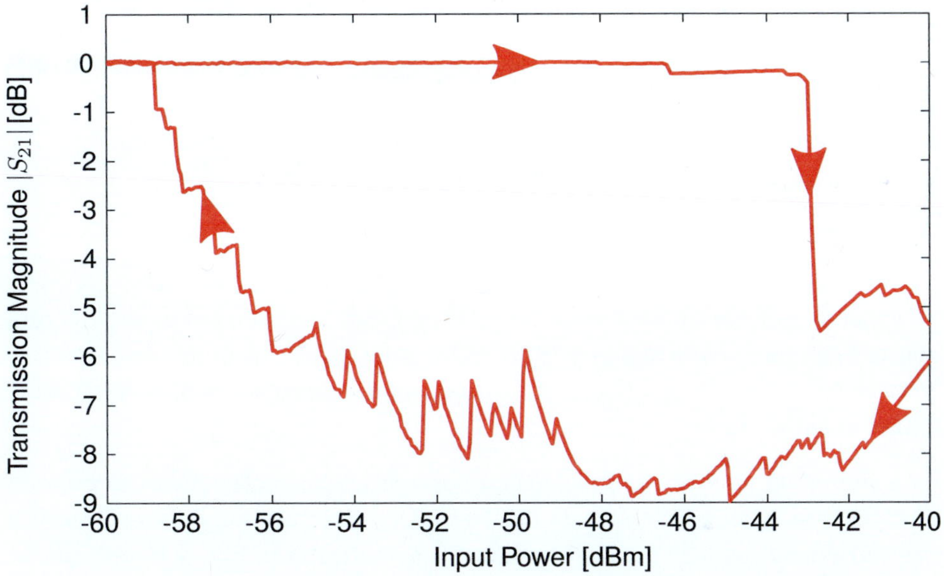

Figure 4.18.: Transmission magnitude measurement for a CPW loaded with 54 SQUID meta-atoms (G2-54). The two curves show how transmission is different for power up- and down-sweeps. The direction of the sweeps is indicated by the arrows. The data have been base line corrected.

Fig. 4.18 shows such a hysteresis curve for sample G2-54 (cf. Sec. 3.1). Depending on the drive frequency, changes in the transmission of up to about 9 dB are observed experimentally corresponding to 89% of the incident power being absorbed and reflected by the array and only 11% transmitted. Apart from a few small jumps in the signal, there is no discernible signature of individual states in the down-sweep curve. On the up-sweep curve, however, a remarkably distinct jump occurs where apparently a significant fraction of the array switches from the transparent to the opaque state.

4.4. Pulse-Switching

This last section is dedicated to a more sophisticated method of probing multi-stability in SQUID meta-atoms. Instead of slowly varying the drive frequency or power, the SQUID can also be forced to go from one state into another by applying microwave pulses.

From the vast variety of possible pulses, I am going to pick out and focus on one of the simplest versions, namely a rectangular envelope microwave (REM) pulse. This scheme allows us to apply some of the knowledge gained from the hysteresis experiments and is also easily implemented experimentally.

4.4.1. Pulse Protocol

In general, the type of pulses that are used here can be generated by multiplying the carrier microwave signal with a rectangular pulse generated by a pulse/-pattern generator. In practice, the multiplication is achieved with a microwave mixer as shown in Fig. 3.10.

In a simplified picture, the mechanism through which such pulses can be used to change the state of the system can be understood from the hysteresis curves introduced earlier. The concept is shown schematically in Fig. 4.19 and can be separated into five steps: First, the system is initialized in a known state at a drive power level low enough so that only one stable state exists. Second, the power is raised to a level at which the SQUID is multi-stable but staying in the small amplitude, transparent state. This is also the point at which the actual measurement is started. Third, after some time, the pulse is applied, increasing the power for a short time, to a level at which the transparent state is no longer stable. Fourth, after the pulse, the driver power is back to the level of step 2 but, as in the hysteresis measurements, the oscillations settle in a high-amplitude, opaque state. Fifth, the measurement is finished and the drive power taken back down to a level outside the multi-stable range.

The lower curve in Fig. 4.19b shows the corresponding transmission expected for such an experiment. In an idealized picture, scattering increases and thus transmission decreases when the meta-atom switches from the transparent to the opaque state.

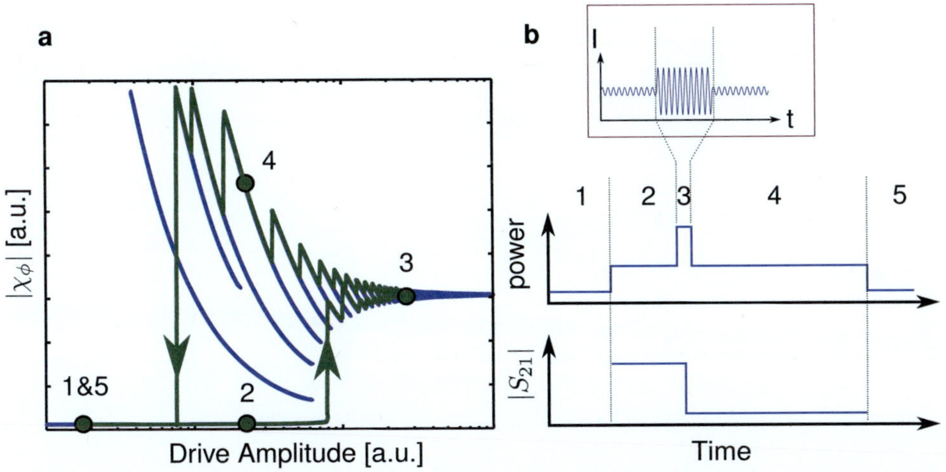

Figure 4.19.: Operating principle of the pulse protocol (sketch). A typical measurement can be divided into five steps, labeled 1 through 5 in both figures. **a**, Magnitude of the oscillations during the five steps (green dots), stable states (blue) and approximate evolution of the system between stable states (green line). **b**, Upper curve: Applied power for the five steps of a typical measurement sequence. The system is initialized during step 1. Data is only taken between steps 2 to 4. The REM pulse itself (step 3) is usually much shorter than depicted here. The inset shows a closeup of the current in the transmission line for a short pulse. Lower curve: Expected transmission magnitude of the CPW. See the explanations in the text for more details.

Although it works well to explain the nature of the experiment, there are a few obvious shortcomings to this simplified picture. Unlike the hysteresis experiment, in which power is swept adiabatically, the changes between power levels in the pulsed experiment happen almost instantaneously on the timescale of the oscillator. Consequently, the green lines in Fig. 4.19a are highly oversimplified, as they imply an adiabatic change of drive amplitude. Furthermore, the pulse duration is usually so small that the system does not reach a steady state during step 3.

From this last remark, it is clear that the final state of the system in step 4 will also depend on the duration of the pulse τ_{pulse} if it is not much longer than the typical relaxation time of the SQUID τ. (For our parameters $\tau \approx 2\,\text{ns}$.)

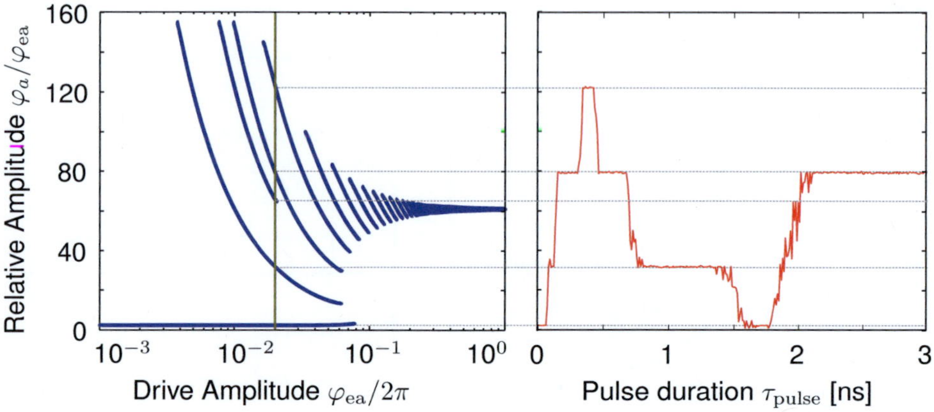

Figure 4.20.: Left: Theoretical stable amplitude values for this particular drive frequency. The vertical green line marks the drive power chosen for the simulation on the right. Right: Simulated dependence of the outcome of the pulse-switching protocol on the pulse duration. Note that the y-axis are identical in both plots. See text for more details.

Just how much influence the pulse duration can have on the outcome is shown in Fig. 4.20. In this simulation, the rf-SQUID equation of motion (Eq. (2.34)) was integrated with a driving term of frequency $f = 1.0075 \cdot f_{r0}$, amplitude $\varphi_{ea} = 4\pi \cdot 10^{-2}$ and Gaussian white noise with standard deviation[7] $\sigma = \pi \cdot 10^{-2} = \varphi_{ea}/4$ for 10 ns. Then a pulse of amplitude 0.3π and varying duration was applied. After that, the integration was continued up to $t = 50$ ns. For each pulse duration value, the simulation was repeated ten times with different initial phase values of the drive. The average of the final amplitude after this procedure is shown relative to φ_{ea} as a function of the pulse duration in the right graph of Fig. 4.20. For comparison, the left plot contains the stable amplitudes expected from Eq. (4.28) in blue and the amplitude of the drive used in the simulation as a vertical green line. The intersections of these lines (that are also shown as horizontal, gray, dashed lines) mark the expected stable amplitude levels in the right plot. There, it can be seen that at least four of the five possible states can be reached depending on the duration of the pulse. Note that from the

[7]This specific noise level was chosen to show that the results are quite robust even for a significant amount of noise (*signal to noise ratio (SNR)* = 4).

hysteresis picture, one would expect the SQUID to end up in the state with $\varphi_{\mathrm{a}}/\varphi_{\mathrm{ea}} \approx 120$ which is only the case for a very narrow range of pulse durations in this simulation.

Fig. 4.21 shows how this situation changes for other drive frequencies and other drive amplitudes. It underlines that, especially for short pulse durations, choosing the "right" pulse to excite the system is by no means a trivial task.

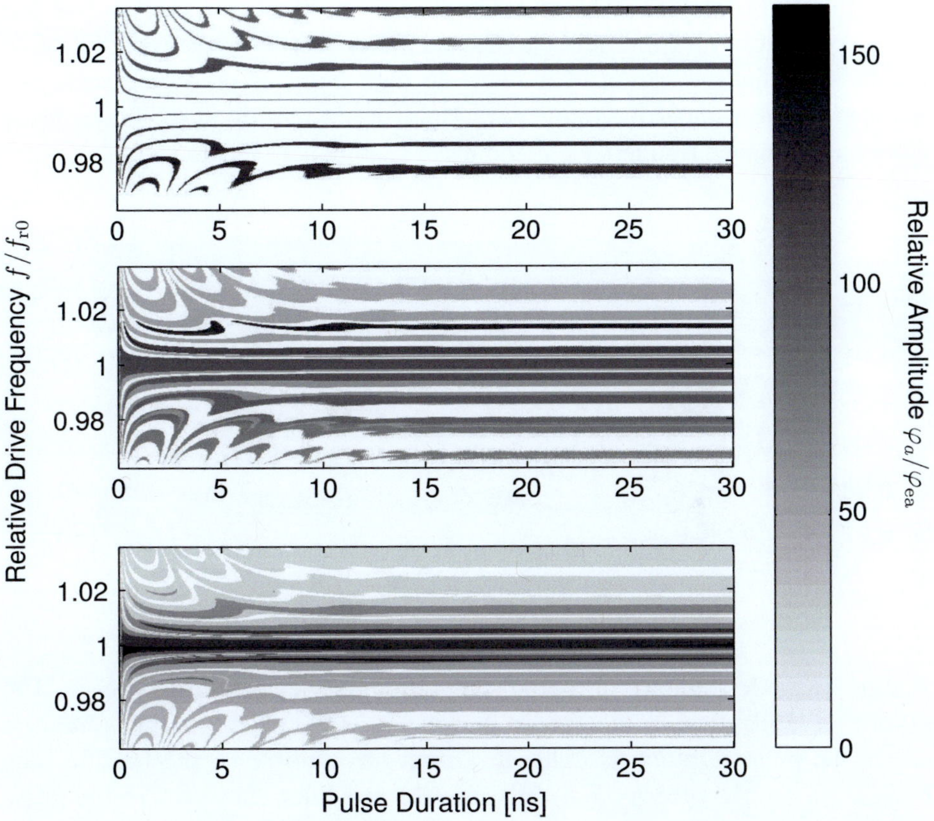

Figure 4.21.: Simulated dependence of the outcome of the pulse-switching protocol as a function of pulse duration and drive frequency for three different values of the drive amplitude. From top to bottom: $\varphi_{\mathrm{ea}} = 0.005 \cdot 2\pi$, $\varphi_{\mathrm{ea}} = 0.01 \cdot 2\pi$, $\varphi_{\mathrm{ea}} = 0.05 \cdot 2\pi$. The final amplitude of the oscillations after an integration time of 80 ns is shown color-coded. The pulse amplitude was $0.5 \cdot 2\pi$.

These results exhibit a strong dependence on the drive frequency even for long pulses but also a particularly intricate interplay between pulse duration and drive frequency for short pulse durations. This can be qualitatively explained as follows. When the SQUID is excited by a pulse with a carrier frequency close to f_{r0}, one can define a beating time τ^* that is inversely proportional to the difference between those frequencies ($\tau^* = 2\pi/|\omega_{r0} - \omega|$). This beating can be seen in the dependence of the switching on the pulse duration and is explicitly highlighted in Fig. 4.22. The effect is most pronounced for short pulses and fades completely for long pulses as the oscillator reaches a stationary state. It can be seen that the fading of the beating lines happens on a time scale that is independent of the drive frequency. Thus, the further the drive frequency is from the resonance frequency f_{r0}, the more beating periods are visible.

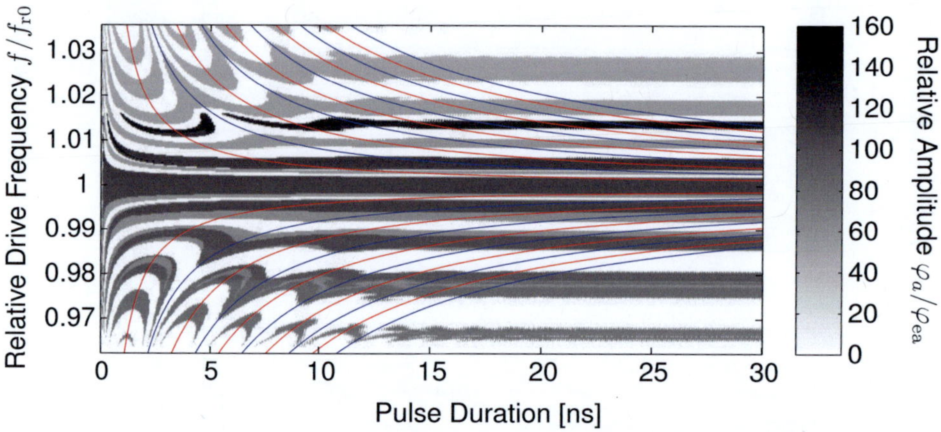

Figure 4.22.: Dependence of the pulse duration on the switching process. The color plot is identical to the central plot in Fig. 4.21. The blue lines indicate where the pulse duration is equal to an integer multiple of the beating time ($\tau_{\text{pulse}} = n \cdot \tau^*$, with $n = 1..5$) while the red lines show where the pulse duration was just half a beating time $\tau_{\text{pulse}} = (n - 1/2) \cdot \tau^*$.

There are several conclusions that can be drawn from these simulations. First, the assumption that the pulse-excitation can be described by the hysteresis as was done in Fig. 4.19 is not entirely justified. Second, using REM pulses, the pulse duration may be used as a parameter to switch to a specific state in a controlled way due to the beating between between drive frequency and self-resonance frequency of the meta atom. Third, exciting the system to a higher state in a wider range of frequencies or drive amplitude levels would require careful adjustment of the pulse properties (i.e. duration and amplitude). Thus, if the goal is to achieve switching over a wide parameter range without adjusting the pulse parameters, REM pulses may not be the best choice. This last point will be picked up again in section 4.4.4, where I will discuss a similar experiment using Gaussian pulses.

4.4.2. Two SQUID Pulse Switching

Using the setup shown in section 3.3 and the method outlined in the last subsection, switching between states by applying pulses was realized. Fig. 4.23a shows an example of single time trace of the transmission through a CPW loaded with two SQUIDs. The SQUID was biased in the multi-stable regime and a pulse was applied after 200 ms biasing time which caused the transmission to drop.

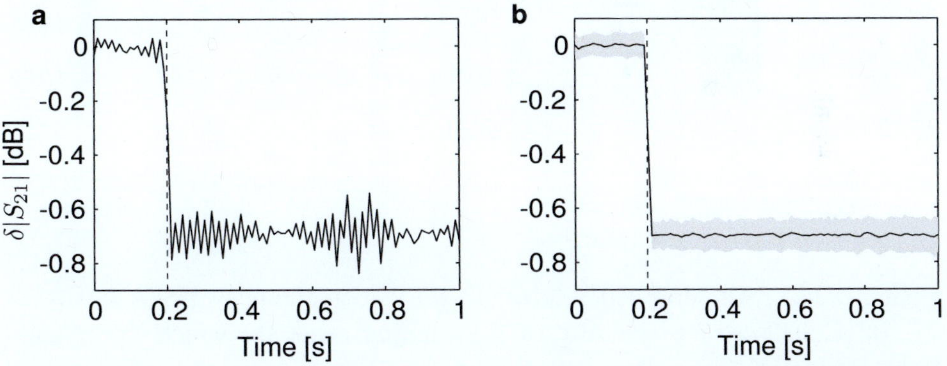

Figure 4.23.: Pulsed measurements on two SQUIDs (G2-2). The y-axis in both plots shows the transmission magnitude relative to the value at the beginning of each trace. **a**, Single time trace of a pulse experiment. A 200 ns pulse was applied after 0.2 s (as indicated by the dashed, vertical line). **b**, Averaged curve over 100 traces like the one on the left. The black line is the mean value, the shaded area marks the corresponding region of one standard deviation $\pm\sigma$ around it.

A rough calibration of the mixer allows us to find the region of multi-stability without difficulties. However, calibration in this setup is a far more delicate task and the characteristics of the mixer (especially its isolation) have been found to also depend on its operating state. As expected, one can identify regions in the parameter space for which switching occurs. In frequency, the effect is most pronounced close to f_{r0}. In drive power, switching only occurs for a certain range of values which roughly corresponds to the region of multi-stability seen in the corresponding hysteresis measurements (cf. Fig. 4.15). At the same time, the pulse power has to exceed a threshold. This dependence can also be seen in Fig. 4.24a, which outlines the relevant parameter range. It shows the probability of a switching event that changes transmission magnitude by more than 0.5 dB as a function of drive power and pulse power. From Fig. 4.24b, one can see that the results also exhibit a dependence on the duration of the pulse for time for surprisingly long time scales up to 25 ns. This is more than one would expect from the single SQUID simulations shown in Fig. 4.21.

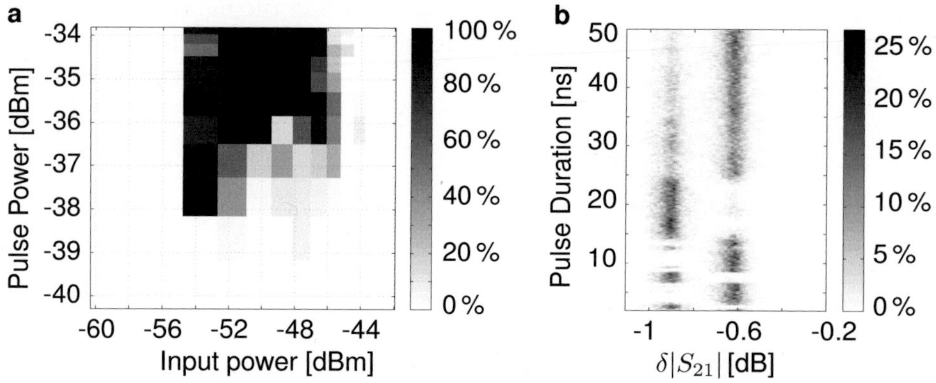

Figure 4.24.: Switching behavior of the CPW containing two SQUIDs (G2-2). **a**, Color-coded probability of a switching event for which the magnitude changes by more than 0.5 dB as a function of drive and pulse power. The pulse duration is constant (200 ns) as is the drive frequency (12.09 GHz $\approx 1.0092 \cdot f_{r0}$). **b**, 2D switching histogram showing the change in transmission magnitude ($\delta|S_{21}|$) as a function of the pulse duration. Each horizontal line is a histogram with 100 bins created from 100 events. The color shows the relative population. Drive and pulse power were constant at -50.6 dBm and -33.8 dBm, respectively.

4.4.3. Pulse Switching of a SQUID Array

Using pulses to switch the state of the 54 SQUID array is analogous to the two-SQUID scenario discussed above. As expected from the hysteresis curve (cf. Fig. 4.18), the magnitude of the change in transmission caused by the pulse is significantly larger than in the case of two SQUIDs. For this particular sample (G2-54), it can reach up to about-9 dB meaning that only about 13% of the incident power is transmitted. Fig. 4.25 contains an example of a single measured trace as well as the statistics over 100 traces, similar to the data shown for two SQUIDs in Fig. 4.23. From the latter, it can be seen that, for this choice of parameters, the transmission can be reduced by 6 dB corresponding to 25% of transmitted incident power.

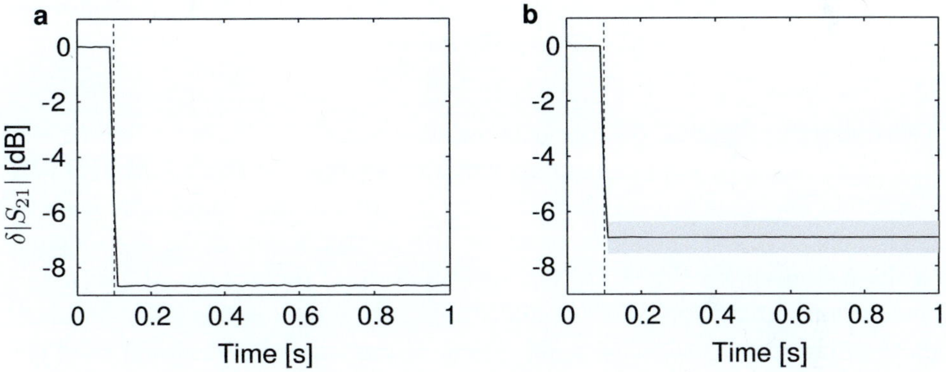

Figure 4.25.: Pulsed measurements on 54 SQUIDs (G2-54) at an input power of $-53\,\text{dBm}$ at $12.09\,\text{GHz} \approx 1.009 \cdot f_{r0}$ and a pulse amplitude of $-38\,\text{dBm}$. The y-axis in both plots shows the transmission magnitude normalized to the value at the beginning of each trace. A pulse of duration $\tau_{\text{pulse}} = 200\,\text{ns}$ was applied after 0.1 s (as indicated by the dashed, vertical line). **a**, Single time trace of the measurement with the largest change in transmission. **b**, Averaged curve over 100 traces like the one on the left. The black line is the mean value, the shaded area marks the corresponding region of one standard deviation $\pm\sigma$ around it.

Due to the large number of possible states in the array, the outcome of each time trace is less predictable than in case of two SQUIDs even for long pulse durations. Sometimes, the state of the array even changes significantly after the pulse has been applied.

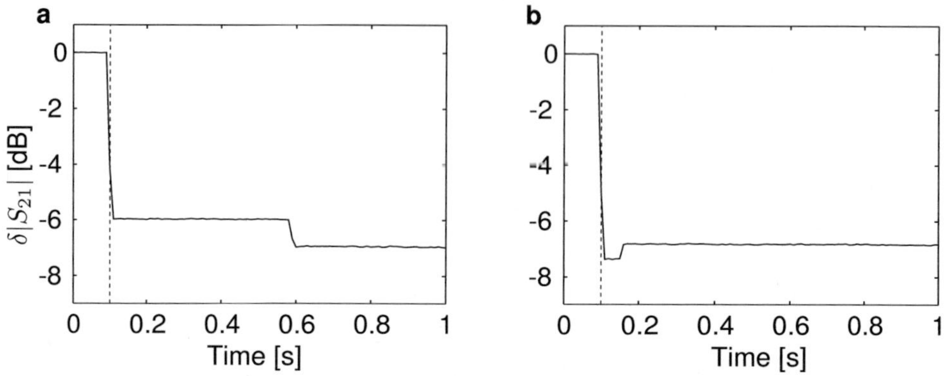

Figure 4.26.: Time traces from the dataset also used in Fig. 4.25. Both these traces show switching after the pulse (indicated by the dashed, vertical line) was applied. **a**, Transmission drops after half the measurement. **b**, Transmission jumps back up shortly after applying the pulse.

Two traces that illustrate this behavior are shown in Fig. 4.26. In the left trace of this figure, it can be seen that the transmission changes by about 1 dB after half a second. This duration is extremely long in terms of the typical time scale of the system as it corresponds to about six billion oscillations of the drive signal. In the data used for Fig. 4.25, such events are rare (only 3 out of 100 traces show a noticeable jump). For larger drive powers, however, this effect becomes more prominent. Qualitatively, this phenomenon can be understood from the generally large number of possible states the array can occupy for a given drive. Even in a simplified picture, the number of states each oscillator can occupy (which factors into the total number of states) depends on the actual state of all oscillators in the chain before it. If we now imagine that one of the SQUIDs in the beginning of the chain switches from one state to another (which may happen due to noise or a small electrical or mechanical disturbance), the rest of the chain has to readjust, which may result in a different magnitude of the transmitted power.

When changing only the drive power for a fixed frequency and the type of pulse, the results of the experiment can be compared to the hysteresis curve as shown in Fig. 4.27. Following the idea outlined in Fig. 4.19, the magnitude of the jump should be given by the difference between the upper and the lower branch of the hysteresis curve. In Fig. 4.27, one can see that, for driving powers below about -51 dB, the magnitude of the switching indeed resembles the difference between the branches of the hysteresis. For higher power levels, this description does not work well any more. From our earlier analysis, this is not surprising. Even more so than in the simpler single SQUID case, the hysteresis only poorly reflects the complex switching dynamics of the SQUID array.

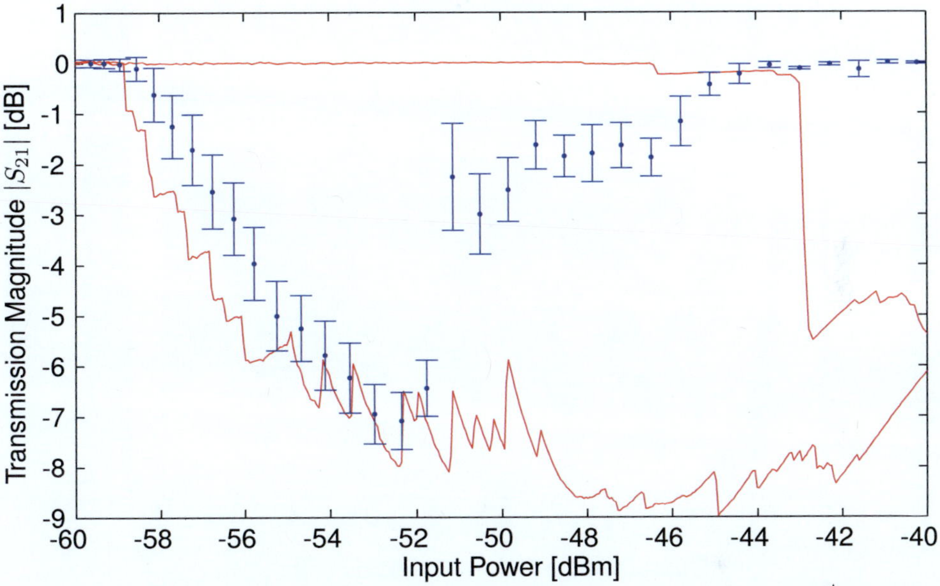

Figure 4.27.: Switching compared to hysteresis in a CPW containing 54 SQUIDs (G2-54). The blue dots and error bars show the mean switching magnitude and standard deviation of 100 samples for different values of the drive power. All other parameters are identical to the ones given in the caption of Fig. 4.25. The red hysteresis curve is shown for comparison.

4.4.4. Alternative Excitation Schemes

The type of pulses used throughout this section were mainly chosen for practical, experimental reasons. They can easily be generated using comparatively simple equipment and do not so easily get deformed by the setup since they are dominated by a single carrier frequency.

Although it was shown that they can be used to switch the SQUID from one state to another, there is a nontrivial dependence of the result on parameters such as the drive frequency (e.g. as shown in Fig. 4.21). This makes it necessary to calibrate the pulses in all measurements in order to achieve switching with a high fidelity.

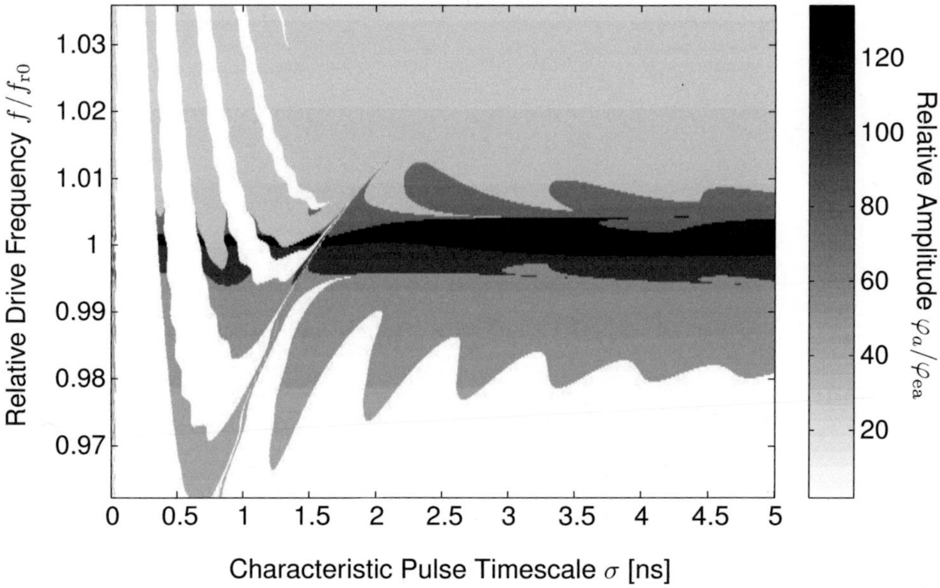

Figure 4.28.: Simulated dependence of the final amplitude after applying a Gaussian pulse of variable width (characterized by σ) as a function of frequency around f_{r0}. The pulses of amplitude $\varphi_p = \pi$ were centered around $t_0 = 10\,\mathrm{ns} + 3\sigma$ and the total integration time was 80 ns. The data can be compared to the central picture in Fig. 4.21, as the same drive amplitude ($\varphi_{ea} = 0.02 \cdot 2\pi$) was used in both simulations.

Using numerical simulations, we can also explore the effect of other types of pulses before actually implementing them in an experiment. Here, I will focus on another basic pulse type, namely a Gaussian pulse, which is added to the harmonic drive signal. The combined driving signal can then be written as

$$\varphi_{ext} = \varphi_{ea}\cos(\omega t) + \varphi_p \exp\left(-\frac{(t-t_0)^2}{2\sigma^2}\right). \tag{4.46}$$

Here, φ_p is the peak amplitude of the pulse that is centered around $t = t_0$ and has a characteristic timescale σ. The width of the pulse is more commonly described by its *full width at half maximum (FWHM)* which is $2\sigma\sqrt{2\ln(2)}$. As the amplitude of the pulse never goes to zero, it is also useful to know that over 99% of the pulse's energy are concentrated between $t = t_0 - 3\sigma$ and $t = t_0 + 3\sigma$.

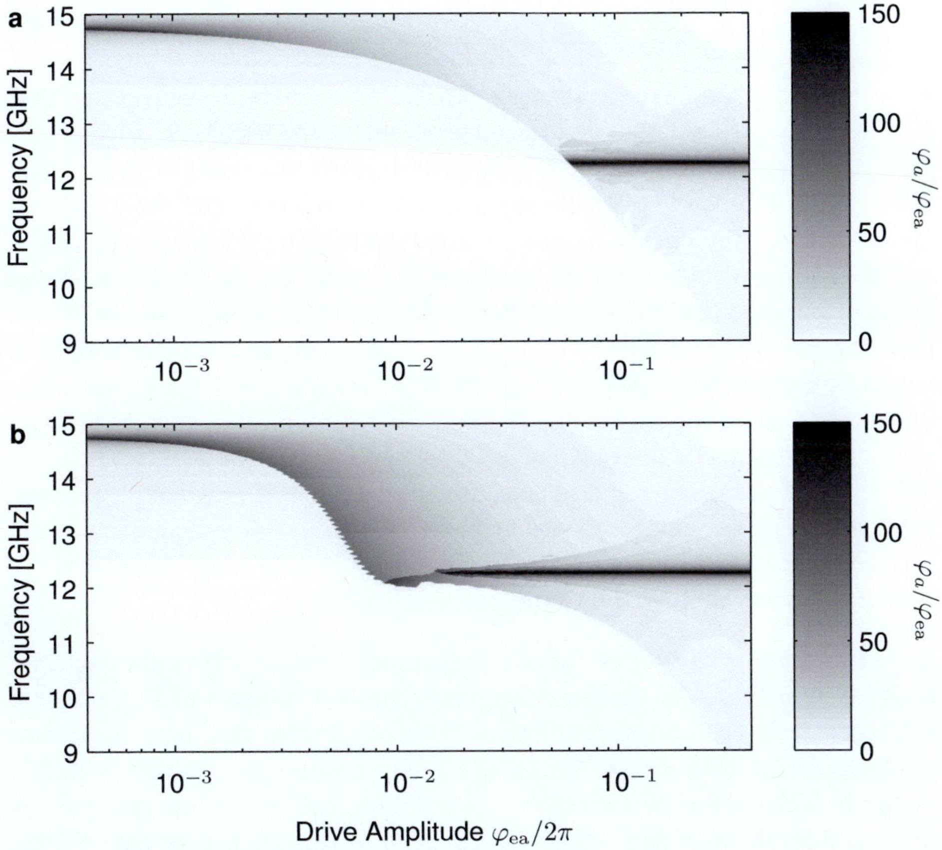

Figure 4.29.: Simulated effect of a Gaussian pulse on the final oscillation amplitude. **a**, Relative amplitude φ_a/φ_{ea} after integrating Eq. (2.34) for 80 ns under a harmonic drive (no pulse) and initial conditions $\varphi(t = 0) = 0$, $\dot{\varphi}(t = 0) = 0$. **b**, In addition to the harmonic drive, a Gaussian pulse with peak amplitude $\varphi_p = \pi$ and width $\sigma = 5$ ns centered at $t_0 = 25$ ns was applied to the system.

When such pulses are applied to the system in a fashion similar to what was shown in the central part of Fig. 4.21, the outcome is quite different from the REM pulses as shown in Fig. 4.28. Most noticeably, the strong oscillations in frequency, that are present even for long pulse durations in Fig. 4.21, are absent for Gaussian pulses. This is not surprising as Gaussian pulses have a much broader spectral dependence than REM pulses. In Fig. 4.28, there is still a significant dependence on the pulse duration especially close to and below the resonance frequency. Above the resonance frequency, however, the results appear to be mostly independent of σ for $\sigma > 2\,\mathrm{ns}$ in this plot.

This knowledge can be used to investigate how the system reacts to such pulses in a wider frequency range and for other values of the harmonic drive amplitude φ_{ea}. Fig. 4.29 contains both a simulation without (a) and with Gaussian pulses (b) as a function of frequency and drive amplitude. The pulse duration and amplitude were chosen to be $\sigma = 5\,\mathrm{ns}$ (FWHM $\approx 11.8\,\mathrm{ns}$), and $\varphi_p = \pi$. Intuitively, one would expect the oscillator to occupy the state with the lowest possible amplitude in Fig. 4.29a as the simulation starts from a "resting" phase particle ($\varphi(t = 0) = 0$, $\dot{\varphi}(t = 0) = 0$). For Fig. 4.29b, on the other hand, one would expect a different picture according to the results shown in Fig. 4.28. The two plots are quite similar for low and high driving amplitudes, but show a significant discrepancy at intermediate values and mainly for frequencies between the weak and strong driving resonance frequencies. In this range of parameters, the Gaussian pulse manages to switch the SQUID from one state into another in a broad range of amplitudes. All of this is consistent with what we know about the occurrence of multi-stability from theory.

To get a clearer picture just for which parameter values the system actually changes its state due to the Gaussian pulse, the two subplots of Fig. 4.29 are subtracted from each other, resulting in Fig. 4.30. In this plot, parameter combinations for which the pulse has no effect have a value close to zero, which is shown in white color. Wherever the color differs from white, the system ends up in a different state depending whether a pulse was applied or not. Darker colors correspond to a larger difference in oscillation amplitude.

The first figure in this chapter (Fig. 4.1) showed the regions of multi-stability as a function of the drive frequency and amplitude. For an ideal pulse, that would switch the SQUID in the whole range of multi-stability, the shape of the colored region in Fig. 4.30 would correspond to the shape seen in the theoretical prediction of Fig. 4.1. This is simply because all dark areas both in Fig. 4.30 and Fig. 4.1 mark regions in the parameter space for which the number of states is larger than one. At first glance, one can see that the general shape of the gray regions is similar in the two figures. Upon closer investigation, however, some discrepancies are visible. For instance, although of similar shape, the main visible feature in Fig. 4.30 is smaller than the region inside the blue dashed line which corresponds to the main multi-stable feature in Fig. 4.1. Also the leaf-like structures that fan out towards higher powers in Fig. 4.1 are only partially present here.

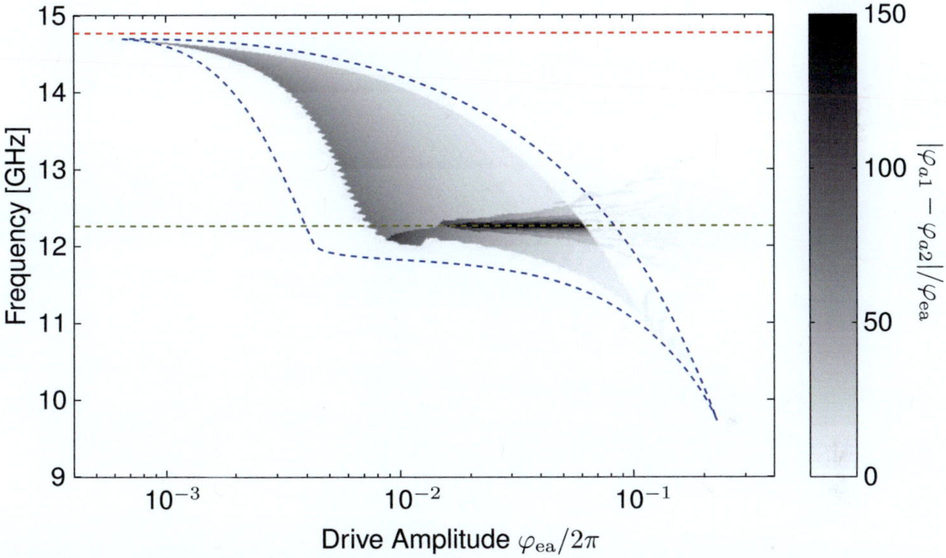

Figure 4.30.: Absolute value of the difference between Fig. 4.29a and Fig. 4.29b. The red and green dashed lines mark the weak and strong driving resonance frequencies, respectively. The area inside the blue dashed line corresponds to the main feature in Fig. 4.1 which showed the theoretical number of stable states under identical conditions.

Apart from the obvious cause of the discrepancy (i.e. that the SQUID is initially in the transparent state and the pulse does not suffice to bring it to an opaque state), there is another effect which is also present here. So far, it was implied that in Fig. 4.29a, the SQUID always settles in the lowest possible state. This assumption, although mostly correct, does not apply for all points in the plot. In fact, due to the initial conditions and the nature of the simulation, the SQUID is already in the opaque state in some regions of Fig. 4.29a. In this case, an additional pulse may not affect the system at all or even bring it back to the transparent state.

Over all, numerical simulations suggest that Gaussian pulses may be a more reliable way of switching the SQUID than the REM pulse used in experiment.

5. Conclusions and Outlook

"There is nothing like looking, if you want to find something. You certainly usually find something, if you look, but it is not always quite the something you were after."

J. R. R. Tolkien in *The Hobbit*

(Spoken by Thorin Oakenshield to the young dwarves.)

When this work was started, the main goal was to build a frequency tunable meta-material from SQUIDs. Although there was also the plan to investigate the nonlinear properties, it was not exactly clear whether nonlinearity could bring any noteworthy advantages to the SQUID as a meta-atom. However, the results of this endeavor clearly show that this is the case, as the observed multi-stability can be used to enhance the functionality of the metamaterial.

In this last part of the thesis, I will summarize the results of the previous chapters and comment on them in the context of current research. Finally, I will give an outlook into a possible future of the topic of this work.

In this thesis, I investigated the nonlinear aspects of the rf-SQUID as a planar, magnetic meta-atom. These meta-atoms are usually resonant structures that constitute the basic building blocks of a metamaterial with the purpose of giving the material extreme (or at least unconventional) magnetic properties. In this context, SQUIDs are often described as a low-loss and tunable alternative to split ring resonators. The low losses come about because of the reduction of Ohmic losses in superconductors while the tunability is an effect of the unique reaction of the SQUID to magnetic field.

This description, however correct, does not do the SQUID meta-atom full justice, as it only covers one of several possible scenarios. In the context of the resistively and capacitively shunted junction model, one can derive an equation of motion for the single-junction SQUID. Its form is that of a nonlinear oscillator with a sinusoidal anharmonicity. Depending on the parameters of the SQUID and the drive, this equation can exhibit all kinds of complex nonlinear behavior and even chaos. For the SQUIDs studied in this thesis, three relevant scenarios can be identified. First, the weak driving limit, in which the SQUID can be treated as a quasi-linear, tunable oscillator. Second, the strong driving limit, where the SQUID effectively behaves as a split-ring resonator with fixed resonance frequency as the effect of the Josephson junction can be neglected. Third, the case of intermediate driving strength, for which the oscillator becomes multi-stable.

The first, weak driving case was experimentally investigated in this thesis on single SQUIDs and one-dimensional SQUID arrays inside coplanar waveguides as well as two-dimensional SQUID arrays in rectangular waveguides. In all three cases, the samples showed a tunable resonance frequency in agreement with theory. Based on the one-dimensional array results, it was shown in Ref. [But+13a] that by loading a coplanar waveguide with such structures, the effective, relative permeability seen by the waveguide can be tuned such that even negative values can be achieved.
The effect of mutual coupling was investigated by comparing two two-dimensional SQUID arrays with different values of coupling strength, each of which contained 441 meta-atoms. As expected from theory, it was found that the coupling influences the resonance frequency of the array. Furthermore, the more strongly coupled of the two arrays showed indications of synchronization. The measurements on two-dimensional arrays are of special interest from the meta-

material point of view as stacks of such arrays could form the basis of a three-dimensional SQUID metamaterial implementation.

The second, strong driving case is of no particular interest here, as the SQUID loses most of its nonlinear properties in this regime. Hence, this case is not discussed in detail in this thesis.

The main focus of this work lay on the third, intermediate driving strength case. This regime is often avoided as it is the least well understood of all three. At the same time, this is the scenario in which the nonlinearity has the strongest impact on the dynamics of the system. Owing to the lack of an analytic theory describing this case, it is usually investigated numerically. In this work, however, a novel analytical model was presented that yields an equation from which the steady state oscillations in the SQUID can be calculated. Although this model is only valid for part of the SQUID parameter space and drive frequencies close to the SQUID resonance frequency, it works for all driving power regimes. Most prominently, it predicts the existence of multi-stability in a broad range of driving powers and frequencies for the SQUIDs that were also used in the weak driving experiments. For these meta-atoms, the model predicts up to ten simultaneously stable states. As amplitude and phase can differ significantly between these states, each of them is characterized by a unique susceptibility to magnetic field. This fact makes the SQUIDs an interesting candidate for a switchable meta-atom.
In direct comparison, the analytical model proved to be in extremely good agreement with numerical simulations for all frequencies of interest to our meta-atoms. Subsequently, the validity of the model could also be confirmed experimentally through measurements on a single SQUID in a coplanar waveguide. In this experiment, the transmission through the waveguide was measured while the power of the driving microwave was swept up and down. The resulting hysteresis curve showed a series of jumps between states that were well described by the analytical model. The same measurement was also performed on a coplanar waveguide sample containing one-dimensional SQUID arrays. As expected for this scenario, the difference between the branches of the observed hysteresis curve was much larger than in the previous case of a single SQUID.

Multi-stability is mainly interesting for the application as a meta-atom if it is possible to switch between the states in situ. Ideally, this should happen in a controllable way and as fast a possible. Although the hysteresis measurements already showed that we are able to change the state in a controlled fashion in situ, the power sweeps employed in this experiment are not particularly fast. To overcome this limitation, another measurement scheme was implemented, in which nanosecond long pulses were applied to the SQUID(s) in addition to the microwave drive. In both numerical simulation and experiment, it was shown that controlled switching of single meta-atoms is possible using rectangular envelope microwave pulses with durations of the order of nanoseconds. Using a waveguide sample that contained 54 rf-SQUIDs, it was demonstrated that the amount of transmitted power can decreased by up to almost 90% by applying a single pulse. This effectively makes the sample an all-optical microwave switch. From these experiments and a subsequent, more detailed numerical analysis, it became clear that there is a non-trivial dependence of the switching fidelity on the parameters of the pulse such as amplitude and duration. As this makes the task of choosing the "right" pulse parameters to switch the meta-atom more complicated, another pulse shape was investigated numerically. From this analysis, it was found that Gaussian pulses should be even better suited to achieve a high switching fidelity than the rectangular envelope microwave pulses studied experimentally.

There are still a number of open questions that will have to be answered in the future. This present thesis has shown that the dynamics of non-hysteretic rf-SQUIDs under microwave drive have not been studied sufficiently in the past. While the presented theory covers the case of zero dc magnetic flux, the question of how the situation changes as a function of this quantity remains open for now. The theory should also be extended to support two-tone or multi-tone excitation signals which would allow us to describe parametric effects in the meta-atom without the usual limitation to weak driving signals.

The experiments and simulations on pulse-switching have so far only scratched the surface of what is possible. From the direct comparison between rectangular envelope microwave pulses and Gaussian pulses, it is clear that the switching fidelity can be improved by optimizing the pulse shape. As similar efforts to optimize pulse shapes for the control of qubits [Mot+09] have lead to a significant increase in gate fidelity, there is a chance that analogous methods could also be applied to our classical system. As all pulsed measurements so far were performed using rectangular envelope microwave pulses, the numerical results on Gaussian pulses will also have to be verified experimentally.

Apart from all these topics, there remains the possibility of using hysteretic rf-

SQUIDs for the same purpose. This idea [DCL06] was in fact proposed before non-hysteretic SQUIDs were considered as meta-atoms. In addition to the dynamic multi-stability discussed in this thesis, they exhibit a static multi-stability that could be harnessed to build another type of multi-stable (and possibly pulse-switchable) metamaterial.

In addition to those points that concern the dynamics of single SQUID meta-atoms, a major field for future research is the investigation of large SQUID arrays. It was shown in this thesis that many of the properties of single SQUIDs (like the ability to tune the resonance by magnetic flux or switch between oscillatory states using pulses) can be extended to arrays if all elements in the array are biased homogeneously and the coupling between them is weak. If one of these two criteria is not fulfilled, however, new effects arise that influence the response of the whole system. This circumstance suggests that an intentionally inhomogeneous bias (such as a magnetic field gradient) can be used as a means to modify the local properties of a SQUID meta-surface or metamaterial. An electromagnetic wave propagating through such a medium could be steered, focused, or otherwise manipulated by adjusting the shape of the bias field gradient. Preliminary studies on strongly coupled arrays that are subjected to an inhomogeneous bias also indicate the presence of synchronized clusters in the array. This idea bears some similarity to so called chimera states that were recently found in numerical simulations of one-dimensional SQUID arrays [LNT14].

Parametric effects that occur for single SQUID meta-atoms might also be harnessed in arrays, where the effect can be strongly enhanced. By optimizing the sample design, it may be possible to realize a form of "wireless" parametric amplification, where gain at a probe frequency can be achieved by pumping the system with a second tone at a different frequency. From a metamaterial perspective, this effect of parametric gain may be useful to overcome the limitation of damping in the artificial medium that is otherwise inherently limiting the performance of metamaterials based on resonant meta-atoms.

From what was shown in this thesis, I hope that I was able to convey my conviction that the rf-SQUID is a very interesting and highly versatile version of a nonlinear meta-atom. I demonstrated frequency tunability by magnetic field and multi-stability at intermediate driving power levels. Although its application is limited to low temperatures, a number of potential applications can be envisioned, especially in the context of superconducting electronics and solid-state quantum computing. The switching in power of almost 90% that was observed in the pulsed experiment, for example, was achieved on a sample that was designed for a different experiment. If improved further, a similar device could serve as an on-chip switch in conjunction with solid-state qubits fabricated using the same technology.

There are also ongoing efforts to build three-dimensional SQUID metamaterials which would have highly magnetic field sensitive transmission properties at microwave frequencies. As with other metamaterials, there is also a general tendency to achieve higher operating frequencies. Using the current Nb based design, it should be possible to implement SQUID metamaterials in the upper GHz or lower THz band, which would open up a new field of applications in communication technologies such as terahertz electronics. Improvements in the high T_c junction technology may also contribute to this development by making SQUIDs from such materials attractive as meta-atoms.

Finally, I would like to point out that some of the concepts outlined in this thesis, such as pulse-switching, may also be applied to non-superconducting meta-atoms. Although the specific type of nonlinearity makes the SQUID unique, phenomena like bistability can also be found in other nonlinear meta-atoms and might be used in a similar fashion.

Figure 5.1.: *The Metamaterial Wall.* An observer uses a continuous light source (flashlight) to illuminate an object through a transparent wall made from a pulse-switchable metamaterial. A sudden strong illumination (camera flash) turns the wall opaque and leaves the observer bewildered. This comic illustrates the counter-intuitive notion that a fundamental change in optical properties comes about without a change in the physical properties of the material.

Zusammenfassung

In dieser Dissertation wurden die nichtlinearen Eigenschaften von rf-SQUID Meta-Atomen untersucht. Unter Meta-Atomen versteht man im Allgemeinen die elementaren Bauteile eines künstlichen Materials (Metamaterial), das mit dem Ziel entwickelt wurde, in der Natur nicht vorkommende, optische Eigenschaften zu realisieren. SQUIDs übernehmen dabei die Rolle resonanter, planarer Meta-Atome, die aufgrund ihrer Bauweise primär an die magnetische Komponente des elektromagnetischen Feldes koppeln. Aufgrund einiger, der Supraleitung geschuldeter Eigenschaften, werden SQUIDs oft als verlustarme und stimmbare Alternative zu sogenannten "split-ring"-Resonatoren dargestellt.

Obwohl diese Beschreibung unter gewissen Voraussetzungen angebracht ist, so tut sie doch den vielseitigen und komplexen Eigenschaften des rf-SQUIDs nicht Genüge. Betrachtet man die Dynamik der Ströme und Spannungen im SQUID auf eine allgemeinere Art und Weise, so erhält man eine nichtlineare Differenzialgleichung zweiter Ordnung. In Abhängigkeit der gewählten Parameter können in diesem System aufgrund seiner sinusförmigen Nichtlinearität eine Vielzahl komplexer Phänomene bis hin zu Chaos auftreten. In dieser Arbeit werden nur sogenannte nicht-hysteretische SQUIDs betrachtet, deren statisches Potential nur ein Minimum aufweist. Ihr Verhalten man in dem hier betrachteten Frequenzintervall grob in drei Bereiche unterteilen kann:
Ist die Amplitude des treibenden, alternierenden magnetischen Feldes sehr klein, so verhält sich das SQUID wie oben angedeutet ähnlich einem gewöhnlichen, linearen Oszillator dessen Resonanzfrequenz durch ein zusätzliches, statisches Magnetfeld stimmbar ist. Im entgegengesetzten Fall, bei sehr starkem treibenden magnetischen Feld, ist das SQUID nicht mehr stimmbar. In dem dazwischen liegenden Bereich ist das SQUID multistabil und es lässt sich keine eindeutige Resonanz definieren.

Der zuerst angesprochene Fall kleiner Amplituden wurde in dieser Arbeit sowohl an einzelnen SQUIDs als auch an ein- und zweidimensionalen Anordnungen untersucht. In allen drei Fällen konnte die Stimmbarkeit der Resonanzfrequenz als Funktion des magnetischen Feldes experimentell verifiziert werden und verhielt sich wie von der Theorie vorhergesagt. Basierend auf den hier vorgestellten Ergebnissen an eindimensionalen SQUID-Ketten konnte in Ref. [But+13a] gezeigt werden, dass mit deren Hilfe die effektive, magnetische Permeabilität eines koplanaren Wellenleiters bis hin zu negativen Werten verstimmt werden kann. Der Einfluss der Kopplung zwischen den SQUIDs wurde in zweidimensionalen Anordnungen von SQUIDs näher untersucht. In Übereinstimmung mit den theoretischen Vorhersagen konnte hier gezeigt werden, dass sich sowohl die Resonanzfrequenz als auch die Schärfe der Resonanz in Abhängigkeit von der Kopplungsstärke verändern. Zusätzlich zeigten diese Messungen Merkmale auf, die auf Synchronisation zwischen Untergruppen von SQUIDs hindeuten.

Der zweite angesprochene Fall starker treibender Magnetfelder war für diese Arbeit von wenig Interesse, da das SQUID dort die meisten seiner nichtlinearen Eigenschaften verliert. Folglich wurde er hier nicht ausführlich behandelt.

Das Hauptaugenmerk dieser Arbeit lag auf dem Bereich zwischen den beiden zuvor angesprochenen Grenzfällen. In diesem Fall hat die Nichtlinearität des Josephson-Kontakts den größten Einfluss auf die Dynamik des Systems. Da sich für diesen Bereich keine allgemeine Lösung der Bewegungsgleichung in geschlossener Form aufschreiben lässt, wird er meist numerisch untersucht. In dieser Arbeit hingegen wurde ein analytischer Ansatz vorgestellt, welcher unter Zuhilfenahme einiger Näherungen einen Ausdruck für die stationären Lösungen der Bewegungsgleichung liefert, die mit der Frequenz der treibenden Mikrowelle oszillieren. Obwohl die Anwendbarkeit dieser Methode von den genauen Parametern des SQUIDs und der Frequenz des treibenden Feldes abhängt, so gilt sie immerhin für alle möglichen treibenden Amplituden und deckt somit alle drei angesprochenen Fälle gleichzeitig ab. Aus diesem Ansatz lässt sich eine präzise Vorhersage für die Streuung von Mikrowellen am SQUID in einem Wellenleiter ableiten, deren markanteste Eigenschaft die Existenz von Multistabilität ist. Für die Parameter der zur Verfügung stehenden SQUIDs sagt das Modell für den mittleren Leistungsbereich bis zu zehn gleichzeitig auftretende stabile Zustände voraus, die sich in Amplitude und Phase stark voneinander unterscheiden. Betrachtet man das SQUID als Meta-Atom, so kann man jedem dieser stationären Zustände eine komplexe magnetische Suszeptibilität zuordnen.
Im direkten Vergleich mit Simulationsergebnissen zeigt sich, dass das Modell bei allen hier betrachteten Frequenzen ausgezeichnete Vorhersagen liefert. Weiterhin konnten die Vorhersagen auch experimentell verifiziert werden. Zu diesem Zwecke wurde die Transmission durch einen Wellenleiter mit eingebettetem SQUID gemessen während die Leistung erst langsam erhöht, dann zurückgefahren wurde. Das Ergebnis war eine Hysterese-Kurve, die mit mehreren Sprüngen einer Vielzahl der vorhergesagten, stabilen Zuständen folgte. In Messungen an eindimensionalen SQUID-Ketten konnten ähnliche Ergebnisse erzielt

werden. Wie erwartet war der Effekt in diesem Fall deutlich ausgeprägter als bei den Versuchen am einzelnen SQUID.

In der Anwendung als Meta-Atom ist Multistabilität insbesondere dann interessant, wenn man zwischen den einzelnen Zuständen in situ und möglichst schnell umschalten kann. Obwohl mit der Hysterese bereits prinzipiell gezeigt wurde, dass sich einzelne stabile Zustände gezielt ansteuern lassen, so verläuft dieser Vorgang nicht sonderlich schnell. Um diese Einschränkung zu umgehen wurde in dieser Arbeit auch die Möglichkeit untersucht, zwischen den Zuständen mittels kurzer Mikrowellenpulse umzuschalten. In Simulation und Experiment konnte gezeigt werden, dass dies an einzelnen SQUIDs möglich ist. Die dafür verwendeten Mikrowellenpulse hatten eine rechteckige Einhüllende und eine Dauer im Bereich von einigen bis einigen hundert Nanosekunden. An SQUID-Ketten konnte auf die gleiche Art gezeigt werden, dass sich die transmittierte Mikrowellenleistung durch den Wellenleiter, in dem sich die SQUIDs befanden, durch einen einzigen Puls um fast 90% verringern ließ. Dieses Verhalten macht solche Systeme als Puls-steuerbare Mikrowellenschalter interessant.
Genauere experimentelle und numerische Untersuchungen ergaben jedoch, dass die verwendete Sorte von Pulsen eine starke Abhängigkeit der Ergebnisse von der gewählten Frequenz zur Folge hatte. Da diese Abhängigkeit die Kalibrierung der Pulse verkompliziert, wurden die numerischen Untersuchungen auf Gauß'sche Pulse ausgeweitet. Die Ergebnisse dieser Simulationen zeigten deutlich, dass Gauß-förmige Pulse besser geeignet sind um SQUIDs zuverlässig über einen breiteren Frequenzbereich zu schalten.

Es bleiben noch etliche offene Fragen, die weitere Forschung auf diesem Gebiet nötig machen. Einerseits gilt es, die gewonnenen theoretischen Methoden so zu erweitern, dass sie auf einen größeren Parameter- und Lösungsbereich anwendbar sind. Andererseits gibt es auch bei der experimentellen Implementierung noch viel Optimierungspotenzial. Abgesehen von der experimentellen Verifikation der numerisch gewonnenen Erkenntnisse gilt es hier insbesondere Bauelemente zu entwickeln, welche die Praxistauglichkeit der hier untersuchten Technologie verdeutlichen. Neben der schon genannten Schaltbarkeit ist hiermit auch der Themenbereich der parametrischen Verstärkung gemeint, der in dieser Arbeit nicht untersucht wurde.
In stark gekoppelten ein- und zweidimensionalen Anordnungen von SQUIDs gilt es die Kopplungseffekte besser zu verstehen und kürzlich vorhergesagte Synchronisationseffekte [LNT14] experimentell zu verifizieren.

Weitere natürliche Entwicklungen für Metamaterialen sind der Schritt von zwei- hin zu dreidimensionalen Anordnungen, sowie der Übergang hin zu höheren Frequenzen. Letzteres könnte durch fortschreitende Verbesserungen von Josephson-Kontakten in Hochtemperatursupraleitern vorangetrieben werden.

Abschließend möchte ich hervorheben, dass das rf-SQUID ein extrem vielseitiger und vielversprechender Kandidat für ein magnetisches Meta-Atom ist, dem wir auch gut 50 Jahre nach seiner Entwicklung immer noch neue Eigenschaften abgewinnen können.

Appendix

A. Detailed Calculations

A.1. RWA: Complex Fourier Coefficients

The complex Fourier coefficients c_1 and c_{-1} from Eq. (4.10) can also be written as

$$c_1 = \frac{1}{2}(a_1 - ib_1), \tag{A.1}$$

$$c_{-1} = c_1^* = \frac{1}{2}(a_1 + ib_1), \tag{A.2}$$

where a_1 and b_1 are the corresponding real Fourier coefficients.

$$a_1 = \frac{2}{T}\int_{t_0}^{t_0+T} \sin(a\cos(\omega t + \delta))\cos(\omega t)dt \tag{A.3}$$

$$b_1 = \frac{2}{T}\int_{t_0}^{t_0+T} \sin(a\cos(\omega t + \delta))\sin(\omega t)dt \tag{A.4}$$

To solve the integrals, we substitute $\omega t + \delta \rightarrow \gamma$ and choose $t_0 = -\delta/\omega$

$$\begin{aligned} a_1 &= \frac{1}{\pi}\int\limits_0^{2\pi} \sin(a\cos(\gamma))\cos(\gamma-\delta)d\gamma && \text{(A.5)} \\ &= \frac{\cos\delta}{\pi}\underbrace{\int\limits_0^{2\pi} \sin(a\cos(\gamma))\cos(\gamma)d\gamma}_{=2\pi J_1(a)} && \text{(A.6)} \\ &\quad + \frac{\sin\delta}{\pi}\underbrace{\int\limits_0^{2\pi} \sin(a\cos(\gamma))\sin(\gamma)d\gamma}_{=0} \\ &= 2\cos(\delta)J_1(a) && \text{(A.7)} \end{aligned}$$

$$\begin{aligned} b_1 &= \frac{1}{\pi}\int\limits_0^{2\pi} \sin(a\cos(\gamma))\sin(\gamma-\delta)d\gamma && \text{(A.8)} \\ &= \frac{\cos\delta}{\pi}\underbrace{\int\limits_0^{2\pi} \sin(a\cos(\gamma))\sin(\gamma)d\gamma}_{=0} && \text{(A.9)} \\ &\quad - \frac{\sin\delta}{\pi}\underbrace{\int\limits_0^{2\pi} \sin(a\cos(\gamma))\cos(\gamma)d\gamma}_{=2\pi J_1(a)} \\ &= -2\sin(\delta)J_1(a) && \text{(A.10)} \end{aligned}$$

The first integrals in Eqns. (A.6) & (A.9) can easily be shown to be zero by shifting the integration boundaries and using the symmetry of the integrand:

$$I_1 = \int_0^{2\pi} \sin(a\cos(\gamma))\sin(\gamma)d\gamma = \int_{-\pi}^{\pi} \sin(a\cos(\gamma))\sin(\gamma)d\gamma \quad (A.11)$$

$$= \int_0^{\pi} \sin(a\cos(\gamma))\sin(\gamma)d\gamma + \int_{-\pi}^{0} \sin(a\cos(\gamma))\sin(\gamma)d\gamma \quad (A.12)$$

$$= 0 \quad (A.13)$$

In the last step we also used $\int_a^b f(x)dx = -\int_b^a f(x)dx$. Using the same approach, the other integrals in Eqns. (A.6) & (A.9) can be related to the Bessel function J_1 according to Eq. (9.54e) in [BS05]:

$$J_1(a) = \frac{2}{\pi}\int_0^{\pi/2} \sin(a\sin(\gamma))\sin(\gamma)d\gamma \quad (A.14)$$

$$= \frac{1}{2\pi}\int_0^{2\pi} \sin(a\cos(\gamma))\cos(\gamma)d\gamma \quad (A.15)$$

So the complex Fourier coefficients c_1 and c_{-1} can be found by inserting Eqns. (A.7) & (A.10) into Eqns. (A.1) & (A.2):

$$c_1 = J_1(a)\,(\cos\delta + i\sin\delta) = J_1(a)e^{i\delta}, \quad (A.16)$$

$$c_{-1} = J_1(a)\,(\cos\delta - i\sin\delta) = J_1(a)e^{-i\delta}. \quad (A.17)$$

B. Technical details

B.1. Methods

B.1.1. Numerical Simulation

Time-dependent simulations of Eq. (2.34) were done in Python using the SciPy.integrate package. The solver itself is the LSODE function of the FORTRAN package odepack. A window of 40 ns was typically chosen for each simulation which is 20 times the RC time-constant of the SQUID and should give it enough time to settle. $\varphi(t)$ was recorded at $2 \cdot 10^4$ discrete points per window. The resulting oversampling factor of about 20 leaves enough room to capture higher harmonics should they occur. A window corresponding to a length of 5 periods of the drive was taken from the end of the full simulation and used to recover amplitude and phase. For this purpose, the data were multiplied with the normalized in-phase and quadrature reference signals and amplitude and phase recovered by averaging.

In case of sweeps (such as the one shown in Fig. 4.6) the system was simulated using manually chosen initial conditions (such as $\varphi = 0$, $\dot{\varphi} = 0$). Then the sweep-parameter (i.e. φ_{ea}) was varied and the simulation repeated but with the last results as new initial conditions. This technique is identical to a single simulation with a time-dependent drive but far more memory-conserving.

To reduce computation time, later simulations (i.e. the ones involving pulses) were done in c++ (odeint library) using GPU acceleration via the *thrust (NVIDIA CUDA)* library.

B.1.2. S-Matrix from Theory (Fig. 4.16)

To calculate the S-Matrix of the loaded CPW in case of a small perturbation (such as a single SQUID), the model outlined in section 3.4 can be used. Although most parameters occurring in this model can be calculated (or at least estimated) from the geometry of the sample and sample holder, some fitting has to be done in order to get a good agreement between theory and experiment. The transmission line parameters in this model (i.e. $l_1, l_2, \beta_1, \beta_2, L_{\mathrm{tl}}, C_{\mathrm{tl}}$ and Z_0) are known or can be calculated analytically from the geometry and verified using finite element field simulation and are thus assumed to be known. The remaining two parameters (i.e. the bond-wire inductance L_{bond} and the mutual inductance M between SQUID and transmission line) are determined from the measured data (cf. Fig. 4.16) using the following simple method: At frequencies close to f_{r0}, the transmission in the weak driving regime is almost independent of M. Consequently, this (constant) value can be used to determine L_{bond}. Subsequently, the difference between transmission magnitude in the strong- and weak-driving regime is used to determine M. The resulting values used for this plot ($M = 6.6\,\mathrm{pF}$, $L_{\mathrm{bond}} = 0.406\,\mathrm{nH}$) are of the same order of magnitude as the a priori calculations ($M = 3.7\,\mathrm{pH}$, $L_{\mathrm{bond}} \approx 0.15 - 0.5$ nH). This discrepancy may very well be a result of the over-simplified model used to calculate the S-matrix.

B.1.3. Retrieving SQUID Parameters From rf Measurements

To retrieve the parameters of the SQUID when only rf measurements are possible (e.g. in case of two-dimensional SQUID array in the rectangular waveguide) the following method can be used: The geometric inductance of the SQUID is simulated using an electromagnetic field solver such as FastHenry. Due to the simple planar geometries these results are usually quite accurate. If there is a significant coupling between the SQUIDs, it has to be accounted for as an effective contribution to the inductance (cf. Eq. (4.45)). Next the SQUID resonance has to be measured in the strong driving and weak driving regime at zero flux (cf. Fig. 4.11). The capacitance of the SQUID can directly be extracted from the strong driving resonance frequency:

$$C = \frac{1}{L\,(2\pi f_{\mathrm{r0}})^2}. \tag{B.1}$$

The critical current is

$$I_c = \frac{\Phi_0}{2\pi L}\left(\frac{f_{max}^2}{f_{r0}^2} - 1\right). \tag{B.2}$$

The resistance can be estimated from the quality factor Q of the strong driving resonance:

$$R = \frac{2\pi f_{r0} L}{Q}. \tag{B.3}$$

B.2. Instruments

The following instruments were used to measure the presented data:

- Vector network analyzer
  Anritsu *VectorStar MS 4642A*
  10 MHz to 20 GHz, 2-port

- Vector network analyzer (used in hysteresis measurements)
  Agilent *PNA-X N5241A*
  10 MHz to 13.5 GHz, 2-port

- Pulse / Pattern Generator
  Agilent *81130A*
  660 MHz, 2-port

- Cryogenic amplifier
  Low Noise Factory *LNF-LNC6_20A*
  6 GHz to 20 GHz, 31 dB gain

- Bias-Ts
  Marki *BT-0025*
  40 kHz to 25 GHz, 0.8 dB insertion loss

- Microwave mixer
  Marki *M1R0726*
  RF: 7 GHz to 26.5 GHz, LO: 7 GHz to 26.5 GHz, IF: DC to 8 GHz

B.3. Software

The following software was used for the work presented in this thesis:

Sample design

- *KLayout*
  `http://www.klayout.de/`
- Autodesk *Autocad*
  `http://www.autodesk.com/`

Mechanical design

- Autodesk *Inventor*
  `http://http://www.autodesk.com/`

Electromagnetic Simulation

- Fast Field Solvers *FastHenry*
  `http://www.fastfieldsolvers.com`
- CST *Microwave Studio*
  `http://www.cst.com/`

Data acquisition

- *QTLab* (IPython based measurement environment)
  `http://qtlab.sourceforge.net`
  `http://www.ipython.org`
  `http://www.python.org`

Data handling and processing

- MathWorks *MATLAB*
  `http://www.mathworks.com`

Numerical differential equation solving

- *SciPy* in conjunction with *Parallel Python*
  `http://www.scipy.org`
  `http://www.parallelpython.com`
  `http://www.python.org`

- GNU C++ using Nvidia CUDA (pulsed simulations on GPU)
  `http://www.odeint.com`
  `http://www.boost.org`
  `http://thrust.github.io/`
  `https://developer.nvidia.com/cuda-zone`
  `https://gcc.gnu.org/`

This document and the figures therein were created using

- LaTeX
  `http://www.latex-project.org`
- *Inkscape*
  `http://www.inkscape.org`

Bibliography

[Abd+13] B. Abdo, K. Sliwa, L. Frunzio, and M. Devoret "Directional Amplification with a Josephson Circuit". *Physical Review X* **3**, 031001 (2013). (Cit. on p. 6).

[AE08] A. Alù and N. Engheta "Plasmonic and metamaterial cloaking: physical mechanisms and potentials". *Journal of Optics A: Pure and Applied Optics* **10** (9), 093002 (2008). (Cit. on p. 2).

[Anl11] S. M. Anlage "The physics and applications of superconducting metamaterials". *Journal of Optics* **13** (2), 024001 (2011). (Cit. on pp. 3–5).

[BCS57] J. Bardeen, L. N. Cooper, and J. R. Schrieffer "Theory of Superconductivity". *Physical Review* **108**, pp. 1175–1204 (1957). (Cit. on p. 25).

[Boa+11] A. D. Boardman, V. V. Grimalsky, Y. S. Kivshar, S. V. Koshevaya, M. Lapine, N. M. Litchinitser, V. N. Malnev, M. Noginov, Y. G. Rapoport, and V. M. Shalaev "Active and tunable metamaterials". *Laser & Photonics Reviews* **5** (2), pp. 287–307 (2011). (Cit. on p. 3).

[BS05] I. N. Bronstein and K. A. Semendjajew. *Taschenbuch der Mathematik*. Ed. by G. Musiol and H. Mühlig. 6., vollst. berarb. und erg. Aufl. Frankfurt am Main: Deutsch, 2005. ISBN: 3-8171-2006-0; 978-3-8171-2006-2; 3-8171-2016-8; 978-3-8171-2016-1 (cit. on p. 129).

[But+13a] S. Butz, P. Jung, L. V. Filippenko, V. P. Koshelets, and A. V. Ustinov "A one-dimensional tunable magnetic metamaterial". *Optics express* **21** (19), pp. 22540–8 (2013). (Cit. on pp. 6, 86, 114, 120).

[But+13b] S. Butz, P. Jung, L. V. Filippenko, V. P. Koshelets, and A. V. Ustinov "Protecting SQUID metamaterials against stray magnetic fields". *Superconductor Science & Technology* **26** (9), 094003 (2013). (Cit. on p. 86).

[But14] S. Butz. "One-Dimensional Tunable Josephson Metamaterials". PhD thesis. Karlsruher Institut für Technologie (KIT), 2014 (cit. on pp. 18, 40, 60, 86).

[CBL07] M. A. Castellanos-Beltran and K. W. Lehnert "Widely tunable parametric amplifier based on a superconducting quantum interference device array resonator". *Applied Physics Letters* **91** (8), 083509 (2007). (Cit. on p. 6).

[Col91] R. E. Collin. *Field theory of guided waves*. 2. ed. The IEEE Press series on electromagnetic wave theory. Piscataway, NJ: IEEE Pr., 1991. ISBN: 0-87942-237-8; 978-0-87942-237-0 (cit. on p. 16).

[CW08] J. Clarke and F. K. Wilhelm "Superconducting quantum bits". *Nature* **453** (7198), 10311042 (2008). (Cit. on p. 6).

[DCL06] C. Du, H. Chen, and S. Li "Quantum left-handed metamaterial from superconducting quantum-interference devices". *Physical Review B* **74** (11), 113105 (2006). (Cit. on pp. 6, 32, 117).

[DCL08] C. Du, H. Chen, and S. Li "Stable and bistable SQUID metamaterials". *Journal of Physics-Condensed Matter* **20** (34), 345220 (2008). (Cit. on p. 6).

[Din+12] Y. Ding, C. Xue, Y. Sun, H. Jiang, Y. Li, H. Li, and H. Chen "Subwavelength electromagnetic switch: Bistable wave transmission of side-coupling nonlinear meta-atom". *Optics Express* **20** (22), pp. 24813–24818 (2012). (Cit. on p. 7).

[DS13] M. H. Devoret and R. J. Schoelkopf "Superconducting Circuits for Quantum Information: An Outlook". *Science* **339** (6124), pp. 1169–1174 (2013). (Cit. on p. 6).

[Eng+13] S. Engelbrecht, F. Kraushofer, A. M. Shuvaev, B. Jin, S. A. Kuznetsov, and A. Pimenov "Asymmetric split-ring resonators: a way toward high-quality metamaterials". *Optical Engineering* **53** (3), pp. 031207–031207 (2013). (Cit. on p. 5).

[Far+08] M. Farhat, S. Enoch, S. Guenneau, and A. B. Movchan "Broadband Cylindrical Acoustic Cloak for Linear Surface Waves in a Fluid". *Physical Review Letters* **101**, 134501 (2008). (Cit. on p. 3).

[FBK83] K. Fesser, A. R. Bishop, and P. Kumar "Chaos in rf SQUIDs". *Applied Physics Letters* **43** (1), pp. 123–124 (1983). (Cit. on p. 77).

[Gu+10] J. Gu, R. Singh, Z. Tian, W. Cao, Q. Xing, M. He, J. W. Zhang, J. Han, H.-T. Chen, and W. Zhang "Terahertz superconductor metamaterial". *Applied Physics Letters* **97** (7), 071102 (2010). (Cit. on p. 5).

[Jac06] J. D. Jackson. *Klassische Elektrodynamik : [7 Tabellen]*. 4., berarb. Aufl. Berlin [u.a.]: de Gruyter, 2006 (cit. on p. 12).

[Jin+10] B. Jin, C. Zhang, S. Engelbrecht, A. Pimenov, J. Wu, Q. Xu, C. Cao, J. Chen, W. Xu, L. Kang, and P. Wu "Low loss and magnetic field-tunable superconducting terahertz metamaterial". *Optics Express* **18** (16), pp. 17504–17509 (2010). (Cit. on p. 5).

[JUA14] P. Jung, A. V. Ustinov, and S. M. Anlage "Progress in superconducting metamaterials". *Superconductor Science and Technology* **27** (7), 073001 (2014). (Cit. on pp. 4, 7).

[Jun+13] P. Jung, S. Butz, S. V. Shitov, and A. V. Ustinov "Low-loss tunable metamaterials using superconducting circuits with Josephson junctions". *Applied Physics Letters* **102** (6), 062601 (2013). (Cit. on pp. 6, 83).

[Jun+14] P. Jung, S. Butz, M. Marthaler, M. V. Fistul, J. Leppäkangas, V. P. Koshelets, and A. V. Ustinov "Multistability and switching in a superconducting metamaterial". *Nature Communications* **5**, 3730 (2014). (Cit. on pp. 7, 66).

[Kau83] R. Kautz "Chaos in Josephson circuits". *IEEE Transactions on Magnetics* **19** (3), pp. 465–474 (1983). (Cit. on p. 77).

[KMM95] A. Kleinsasser, R. Miller, and W. Mallison "Dependence of critical current density on oxygen exposure in Nb-AlO/sub x/-Nb tunnel junctions". *Applied Superconductivity, IEEE Transactions on* **5** (1), pp. 26–30 (1995). (Cit. on p. 40).

[KO11] H. Kamerlingh Onnes "Further experiments with liquid helium. G. On the electrical resistance of pure metals, etc. VI. On the sudden change in the rate at which the resistance of mercury disappears." *Comm. Phys. Lab. Univ. Leiden* **124c**, (1911). (Cit. on p. 25).

[Lik91] K. K. Likharev. *Dynamics of Josephson junctions and circuits*. 2nd pr. Philadelphia, Pa. [u.a.]: Gordon and Breach, 1991. ISBN: 2-88124-042-9 (cit. on p. 25).

[LL12] D. Lu and Z. Liu "Hyperlenses and metalenses for far-field super-resolution imaging". *Nature Communications* **3**, 1205 (2012). (Cit. on p. 3).

[LL98] L. D. Landau and E. M. Lifshitz. *Course of theoretical physics*. 3. ed., repr. Vol. 1: Mechanics. Oxford [u.a.]: Butterworth-Heinemann, 1998. ISBN: 0-7506-2896-0 (cit. on pp. 12, 36, 71).

[LNT14] N Lazarides, G Neofotistos, and G. Tsironis "Chimeras in SQUID Metamaterials". *arXiv:1408.6072*, (2014). (Cit. on pp. 117, 122).

[LSK14] M. Lapine, I. V. Shadrivov, and Y. S. Kivshar "Colloquium: Non-linear metamaterials". *Review of Modern Physics* **86**, pp. 1093–1123 (2014). (Cit. on p. 7).

[LT07] N. Lazarides and G. P. Tsironis "rf superconducting quantum interference device metamaterials". *Applied Physics Letters* **90** (16), 163501 (2007). (Cit. on p. 6).

[LT13] N. Lazarides and G. P. Tsironis "Multistability and self-organization in disordered SQUID metamaterials". *Superconductor Science & Technology* **26** (8), 084006 (2013). (Cit. on pp. 7, 88, 91).

[Mac+13] P. Macha, G. Oelsner, J.-M. Reiner, M. Marthaler, S. André, G. Schön, U. Huebner, H.-G. Meyer, E. Il'ichev, and A. V. Ustinov "Implementation of a Quantum Metamaterial". *arXiv:1309.5268*, (2013). (Cit. on p. 7).

[Mac13] P. Macha. "A path towards quantum metamaterials". PhD thesis. Karlsruher Institut für Technologie (KIT), 2013 (cit. on p. 7).

[Mae+95] M. Maezawa, M. Aoyagi, H. Nakagawa, I. Kurosawa, and S. Takada "Specific capacitance of Nb/AlOx/Nb Josephson junctions with critical current densities in the range of 0.1–18 kA/cm2". *Applied Physics Letters* **66** (16), pp. 2134–2136 (1995). (Cit. on p. 40).

[MG10] A. I. Maimistov and I. R. Gabitov "Nonlinear response of a thin metamaterial film containing Josephson junctions". *Optics Communications* **283** (8), pp. 1633–1639 (2010). (Cit. on p. 6).

[Mot+09] F. Motzoi, J. M. Gambetta, P. Rebentrost, and F. K. Wilhelm "Simple Pulses for Elimination of Leakage in Weakly Nonlinear Qubits". *Physical Review Letters* **103**, p. 110501 (2009). (Cit. on p. 116).

[NB11] G. V. Naik and A. Boltasseva "A comparative study of semiconductor-based plasmonic metamaterials". *Metamaterials* **5** (1), pp. 1–7 (2011). (Cit. on p. 3).

[Ovc+13] E. A. Ovchinnikova, S. Butz, P. Jung, V. P. Koshelets, L. V. Filippenko, A. S. Averkin, S. V. Shitov, and A. V. Ustinov "Design and experimental study of superconducting left-handed transmission lines with tunable dispersion". *Superconductor Science & Technology* **26** (11), 114003 (2013). (Cit. on p. 6).

[Pen+96] J. B. Pendry, A. J. Holden, W. J. Stewart, and I. Youngs "Extremely low frequency plasmons in metallic mesostructures". *Physical Review Letters* **76** (25), pp. 4773–4776 (1996). (Cit. on p. 2).

[Pen+99] J. B. Pendry, A. J. Holden, D. J. Robbins, and W. J. Stewart "Magnetism from conductors and enhanced nonlinear phenomena". *IEEE Transactions on Microwave Theory and Techniques* **47** (11), pp. 2075–2084 (1999). (Cit. on p. 2).

[Pen00] J. B. Pendry "Negative refraction makes a perfect lens". *Physical Review Letters* **85** (18), pp. 3966–3969 (2000). (Cit. on p. 2).

[Poz05] D. M. Pozar. *Microwave engineering*. 3. ed. Hoboken, NJ: Wiley, 2005. ISBN: 0-471-44878-8; 978-0-471-44878-5 (cit. on pp. 16, 19, 24, 58).

[RA06] M. C. Ricci and S. M. Anlage "Single superconducting split-ring resonator electrodynamics". *Applied Physics Letters* **88** (26), 264102 (2006). (Cit. on p. 5).

[RHL77] N. Rouche, P. Habets, and M. Laloy. *Stability theory by Liapunovs direct method.* Applied mathematical sciences ; 22. New York; Heidelberg: Springer, 1977. ISBN: 0-387-90258-9; 3-540-90258-9 (cit. on p. 69).

[Ric+07] M. C. Ricci, H. Xu, R. Prozorov, A. P. Zhuravel, A. V. Ustinov, and S. M. Anlage "Tunability of superconducting metamaterials". *IEEE Transactions on Applied Superconductivity* **17** (2), pp. 918–921 (2007). (Cit. on pp. 5, 19).

[ROA05] M. Ricci, N. Orloff, and S. M. Anlage "Superconducting metamaterials". *Applied Physics Letters* **87** (3), 034102 (2005). (Cit. on pp. 4, 19).

[RS83] R. K. Ritala and M. M. Salomaa "Odd and even subharmonics and chaos in RF SQUIDS". *Journal of Physics C: Solid State Physics* **16** (14), p. L477 (1983). (Cit. on p. 77).

[RS84] R. K. Ritala and M. M. Salomaa "Chaotic dynamics of periodically driven rf superconducting quantum interference devices". *Physical Review B* **29**, pp. 6143–6154 (1984). (Cit. on p. 77).

[Sch+06] D. Schurig, J. J. Mock, B. J. Justice, S. A. Cummer, J. B. Pendry, A. F. Starr, and D. R. Smith "Metamaterial electromagnetic cloak at microwave frequencies". *Science* **314** (5801), pp. 977–980 (2006). (Cit. on p. 2).

[Sch97] V. V. Schmidt. *The physics of superconductors : introduction to fundamentals and applications; with 51 problems with solutions*. Ed. by P. H. Müller. Berlin: Springer, 1997. ISBN: 3-540-61243-2; 978-3-642-08251-1 (cit. on pp. 25, 29).

[She+01] R. A. Shelby, D. R. Smith, S. C. Nemat-Nasser, and S. Schultz "Microwave transmission through a two-dimensional, isotropic, left-handed metamaterial". *Applied Physics Letters* **78** (4), pp. 489–491 (2001). (Cit. on p. 2).

[Sim01] R. N. Simons. *Coplanar waveguide circuits, components, and systems*. Wiley series in microwave and optical engineering. New York, NY: Wiley-Interscience, 2001. ISBN: 0-471-16121-7 (cit. on p. 18).

[Sin+13] R. Singh, D. Roy Chowdhury, J. Xiong, H. Yang, A. K. Azad, A. J. Taylor, Q. X. Jia, and H.-T. Chen "Influence of film thickness in THz active metamaterial devices: A comparison between superconductor and metal split-ring resonators". *Applied Physics Letters* **103** (6), 061117 (2013). (Cit. on p. 5).

[Smi+00] D. R. Smith, W. J. Padilla, D. C. Vier, S. C. Nemat-Nasser, and S. Schultz "Composite medium with simultaneously negative permeability and permittivity". *Physical Review Letters* **84** (18), pp. 4184–4187 (2000). (Cit. on pp. 2, 4).

[SSS01] R. A. Shelby, D. R. Smith, and S. Schultz "Experimental verification of a negative index of refraction". *Science* **292** (5514), pp. 77–79 (2001). (Cit. on pp. 2, 13).

[Tin04] M. Tinkham. *Introduction to superconductivity*. 2. ed. Dover books on physics. Mineola, NY: Dover Publ., 2004. ISBN: 0-486-43503-2; 978-0-486-43503-9 (cit. on p. 25).

[Tre+13] M. Trepanier, D. M. Zhang, O. Mukhanov, and S. M. Anlage "Realization and Modeling of Metamaterials Made of rf Superconducting Quantum-Interference Devices". *Physical Review X* **3** (4), 041029 (2013). (Cit. on pp. 6, 61).

[VDS09] R. Vijay, M. H. Devoret, and I. Siddiqi "Invited Review Article: The Josephson bifurcation amplifier". *Review of Scientific Instruments* **80** (11), 111101 (2009). (Cit. on pp. 7, 65).

[Ves68] V. G. Veselago "THE ELECTRODYNAMICS OF SUBSTANCES WITH SIMULTANEOUSLY NEGATIVE VALUES OF ε AND μ". *Soviet Physics Uspekhi* **10** (4), p. 509 (1968). (Cit. on p. 2).

[Wan+08] B. Wang, J. Zhou, T. Koschny, and C. M. Soukoulis "Nonlinear properties of split-ring resonators". *Optics Express* **16** (20), pp. 16058–16063 (2008). (Cit. on p. 7).

[Wol12] K. Wolff. "Construction of a Cryogenic Microwave Sample Box". Bachelor Thesis. Karlsruhe Institute of Technology, 2012 (cit. on p. 53).

[Wu+11] J. Wu, B. Jin, Y. Xue, C. Zhang, H. Dai, L. Zhang, C. Cao, L. Kang, W. Xu, J. Chen, and P. Wu "Tuning of superconducting niobium nitride terahertz metamaterials". *Optics Express* **19** (13), pp. 12021–12026 (2011). (Cit. on p. 5).

[Zha+12] C. H. Zhang, J. B. Wu, B. B. Jin, Z. M. Ji, L. Kang, W. W. Xu, J. Chen, M. Tonouchi, and P. H. Wu "Low-loss terahertz metamaterial from superconducting niobium nitride films". *Optics Express* **20** (1), pp. 42–47 (2012). (Cit. on p. 5).

List of Publications

1. P. Jung, S. Butz, S. V. Shitov and A. V. Ustinov. "Low-loss Tunable Metamaterials using Superconducting Circuits with Josephson Junctions". *Appl. Phys. Lett.* **102**, 062601 (2013).

2. A. P. Zhuravel, B. G. Ghamsari, C. Kurter, P. Jung, S. Remillard, J. Abrahams, O. Lukashenko, A. V. Ustinov and S. M. Anlage "Imaging the Anisotropic Nonlinear Meissner Effect in Nodal $YBa_2Cu_3O_{7-\delta}$ Thin-Film Superconductors". *Phys. Rev. Lett.* **110**, 087002 (2013).

3. S. Probst, H. Rotzinger, S. Wünsch, P. Jung, M. Jerger, M. Siegel, A. V. Ustinov and P. A. Bushev "Anisotropic Rare-Earth Spin Ensemble Strongly Coupled to a Superconducting Resonator". *Phys. Rev. Lett.* **110**, 157001 (2013).

4. S. Butz, P. Jung, L. V. Filippenko, V. P. Koshelets and A. V. Ustinov. "Protecting SQUID Metamaterials against Stray Magnetic Fields". *Supercond. Sci. Technol.* **26**, 094003 (2013).

5. S. Butz, P. Jung, L. V. Filippenko, V. P. Koshelets and A. V. Ustinov. "A One-Dimensional Tunable Magnetic Metamaterial". *Opt. Express* **21**, 22540 (2013).

6. E. A. Ovchinnikova, S. Butz, P. Jung, V. P. Koshelets, L. V. Filippenko, A. S. Averkin, S. V. Shitov and A. V. Ustinov. "Design and Experimental Study of Superconducting Left-Handed Transmission Lines with Tunable Dispersion". *Supercond. Sci. Technol.* **26**, 114003 (2013).

7. P. Jung, S. Butz, M. Marthaler, M. V. Fistul, J. Leppäkangas, V. P. Koshelets and A. V. Ustinov. "Multistability and Switching in a Superconducting Metamaterial". *Nat. Commun.* **5**, 3730 (2014).

8. N. Maleeva, M. V. Fistul, A. Karpov, A. P. Zhuravel, A. Averkin, P. Jung and A. V. Ustinov "Electrodynamics of a ring-shaped spiral resonator". *J. Appl. Phys.* **115**, 064910 (2014).

9. S. Butz, P. Jung, L. V. Filippenko, V. P. Koshelets and A. V. Ustinov. "A One-Dimensional Tunable Magnetic Metamaterial: Erratum". *Opt. Express* **22**, 13041 (2014).

10. P. Jung, A. V. Ustinov and S. M. Anlage "Progress in Superconducting Metamaterials". *Supercond. Sci. Technol.* **26**, 094003 (2014).

11. S. K. Remillard, D. Kirkendall, G. Ghigo, R. Gerbaldo, L. Gozzelino, F. Laviano, Z. Yang, N. A. Mendelsohn, B. G. Ghamsari, B. Friedman, P. Jung and S. M. Anlage "Microwave nonlinearity and photoresponse of superconducting resonators with columnar defect micro-channels". *Supercond. Sci. Technol.* **27**, 095006 (2014).

Acknowledgements

In this final part of the thesis, I would like to thank all those people that influenced and assisted me during my time as a PhD student at the Physikalisches Institut.

First of all, I want to thank Prof. Alexey V. Ustinov for giving me the chance to work in his group on this exciting project. Throughout all my time here, he has given me the support I needed to be productive and successful while leaving me the freedom to think creatively and pursue my own ideas. Despite his many obligations, he has alway found the time to listen and guide me in the right direction. For all these things, I am deeply grateful.
Likewise, I would like to thank Prof. Steven M. Anlage. Not only has he agreed to be the second referee for my thesis, he has also given me the chance to work in his lab in Marlyland. For all this as well as for the countless discussions and hours together in the lab, I am very thankful.

I owe my gratitude to many people in the group, first and foremost to my friend and colleague Susanne Butz. I thank her for all the support and the good times we had together and for resisting the urge to throw things at me (most of the time). My thanks also go out to all the other people in the group, especially my fellow PhD students. I would like to thank Hannes Rotzinger and Martin Weides for all the advice and the productive discussions as well as Sasha Lukashenko for sharing his expertise and helping me on countless occasions.

I am very grateful to both the mechanics and electronics workshop of the Physikalisches Institut, especially Michael Meyer and Roland Jehle for helping to build the measurement setup. From Roland Jehle, I have learned a great deal concerning the proper design of electrical circuits and PCBs, while Michael Meyer was always willing and able to help me to improve my CAD skills.

Thank you to all the external coworkers and collaborators who worked with me on the projects concerning this thesis including Michael Marthaler and Mikhail

Fistul for the theoretical work, Daimeng Zhang and Melissa Trepanier from the Maryland group, Valery Koshelets who organized the fabrication of the samples and all the members of the MISiS group that were involved in this project.

I thank the *Helmholtz International Research School for Teratronics (HIRST)* for their support and for funding one year of my thesis as well as the *Karlsruhe House of Young Scientists (KHYS)* for supporting my stay at the University of Maryland through their research travel scholarship.

My personal thanks go out to my family, especially my mother and my father, who have always supported me in all my endeavors. I would also like to thank my "old" and my "new" friends, from here and abroad: Thank you Jeff, Daniel, Benny, Sanne (again), Jochen, Tob, Patty and all the others! Finally, I want to thank my girlfriend Christina for giving me all the love and support and for enduring my early-morning physics talk.

Experimental Condensed Matter Physics
(ISSN 2191-9925)

Herausgeber
Physikalisches Institut

Prof. Dr. Hilbert von Löhneysen
Prof. Dr. Alexey Ustinov
Prof. Dr. Georg Weiß
Prof. Dr. Wulf Wulfhekel

Die Bände sind unter www.ksp.kit.edu als PDF frei verfügbar oder als Druckausgabe bestellbar.

Band 1 Alexey Feofanov
Experiments on flux qubits with pi-shifters. 2011
ISBN 978-3-86644-644-1

Band 2 Stefan Schmaus
Spintronics with individual metal-organic molecules. 2011
ISBN 978-3-86644-649-6

Band 3 Marc Müller
Elektrischer Leitwert von magnetostriktiven Dy-Nanokontakten. 2011
ISBN 978-3-86644-726-4

Band 4 Torben Peichl
Einfluss mechanischer Deformation auf atomare Tunnelsysteme – untersucht mit Josephson Phasen-Qubits. 2012
ISBN 978-3-86644-837-7

Band 5 Dominik Stöffler
Herstellung dünner metallischer Brücken durch Elektromigration und Charakterisierung mit Rastersondentechniken. 2012
ISBN 978-3-86644-843-8

Band 6 Tihomir Tomanic
Untersuchung des elektronischen Oberflächenzustands von Ag-Inseln auf supraleitendem Niob (110). 2012
ISBN 978-3-86644-898-8

Band 7 Lukas Gerhard
Magnetoelectric coupling at metal surfaces. 2013
ISBN 978-3-7315-0063-6

Experimental Condensed Matter Physics
(ISSN 2191-9925)

Band 8 Kirill Fedorov
Fluxon readout for superconducting flux qubits. 2013
ISBN 978-3-7315-0067-4

Band 9 Jochen Zimmer
Cooper pair transport in arrays of Josephson junctions. 2014
ISBN 978-3-7315-0130-5

Band 10 Oliver Berg
Elektrischer Transport durch Nanokontakte von Selten-Erd-Metallen. 2014
ISBN 978-3-7315-0209-8

Band 11 Grigorij Jur'evic Grabovskij
Investigation of coherent microscopic defects inside the tunneling barrier of a Josephson junction. 2014
ISBN 978-3-7315-0210-4

Band 12 Cornelius Thiele
STM Characterization of Phenylene-Ethynylene Oligomers on Au(111) and their Integration into Carbon Nanotube Nanogaps. 2014
ISBN 978-3-7315-0235-7

Band 13 Michael Peter Schackert
Scanning Tunneling Spectroscopy on Electron-Boson Interactions in Superconductors. 2014
ISBN 978-3-7315-0238-8

Band 14 Susanne Butz
One-Dimensional Tunable Josephson Metamaterials – Eindimensionale stimmbare Josephson Metamaterialien. 2014
ISBN 978-3-7315-0271-5

Band 15 Philipp Jung
Nonlinear Effects in Superconducting Quantum Interference Meta-Atoms. 2014
ISBN 978-3-7315-0294-4